ASSESSMENT OF CORROSION EDUCATION

Committee on Assessing Corrosion Education

National Materials Advisory Board

Division on Engineering and Physical Sciences

NATIONAL RESEARCH COUNCIL
OF THE NATIONAL ACADEMIES

THE NATIONAL ACADEMIES PRESS
Washington, D.C.
www.nap.edu

THE NATIONAL ACADEMIES PRESS   500 Fifth Street, N.W.   Washington, DC 20001

NOTICE: The project that is the subject of this report was approved by the Governing Board of the National Research Council, whose members are drawn from the councils of the National Academy of Sciences, the National Academy of Engineering, and the Institute of Medicine. The members of the committee responsible for the report were chosen for their special competences and with regard for appropriate balance.

This study was supported by Contract No. FA8501-06-D-0001 between the National Academy of Sciences and the Department of Defense. Any opinions, findings, conclusions, or recommendations expressed in this publication are those of the authors and do not necessarily reflect the views of the organizations or agencies that provided support for the project.

Cover design by Steven Coleman.
Cover Image © Digital Vision/Cocoon.

International Standard Book Number-13:   978-0-309-11703-6
International Standard Book Number-10:   0-309-11703-8
Library of Congress Control Number: 2008944165

Available in limited quantities from

National Materials Advisory Board
500 Fifth Street, N.W.
Washington, DC 20001
nmab@nas.edu
http://www.nationalacademies.edu/nmab

Additional copies of this report are available from the National Academies Press, 500 Fifth Street, N.W., Lockbox 285, Washington, DC 20055; (800) 624-6242 or (202) 334-3313 (in the Washington metropolitan area); Internet, http://www.nap.edu.

THE NATIONAL ACADEMIES
Advisers to the Nation on Science, Engineering, and Medicine

The **National Academy of Sciences** is a private, nonprofit, self-perpetuating society of distinguished scholars engaged in scientific and engineering research, dedicated to the furtherance of science and technology and to their use for the general welfare. Upon the authority of the charter granted to it by the Congress in 1863, the Academy has a mandate that requires it to advise the federal government on scientific and technical matters. Dr. Ralph J. Cicerone is president of the National Academy of Sciences.

The **National Academy of Engineering** was established in 1964, under the charter of the National Academy of Sciences, as a parallel organization of outstanding engineers. It is autonomous in its administration and in the selection of its members, sharing with the National Academy of Sciences the responsibility for advising the federal government. The National Academy of Engineering also sponsors engineering programs aimed at meeting national needs, encourages education and research, and recognizes the superior achievements of engineers. Dr. Charles M. Vest is president of the National Academy of Engineering.

The **Institute of Medicine** was established in 1970 by the National Academy of Sciences to secure the services of eminent members of appropriate professions in the examination of policy matters pertaining to the health of the public. The Institute acts under the responsibility given to the National Academy of Sciences by its congressional charter to be an adviser to the federal government and, upon its own initiative, to identify issues of medical care, research, and education. Dr. Harvey V. Fineberg is president of the Institute of Medicine.

The **National Research Council** was organized by the National Academy of Sciences in 1916 to associate the broad community of science and technology with the Academy's purposes of furthering knowledge and advising the federal government. Functioning in accordance with general policies determined by the Academy, the Council has become the principal operating agency of both the National Academy of Sciences and the National Academy of Engineering in providing services to the government, the public, and the scientific and engineering communities. The Council is administered jointly by both Academies and the Institute of Medicine. Dr. Ralph J. Cicerone and Dr. Charles M. Vest are chair and vice chair, respectively, of the National Research Council.

www.national-academies.org

COMMITTEE ON ASSESSING CORROSION EDUCATION

Preface

The threat from the degradation of materials in the engineered products that drive our economy, keep our citizenry healthy, and keep us safe from terrorism and belligerent threats has been well documented over the years. The cost to the economy, as discussed in this report, is significant. And yet little effort appears to have been made to apply the nation's engineering community to developing a better understanding of corrosion and the mitigation of its effects.

At the direction of the Congress, however, the Department of Defense (DOD) has begun to pay more attention to the threat corrosion poses to the national security infrastructure. As part of that effort Congress instructed DOD to commission a National Research Council (NRC) report to assess the current state of corrosion engineering education in the United States.

Subsequently the Committee on Assessing Corrosion Education was appointed. The committee's charge was as follows:

> A committee of the National Academies will be convened to
> 1. Assess the level and effectiveness of existing engineering curricula in corrosion science and technology, including corrosion prevention and control, and
> 2. Recommend actions that could enhance the corrosion-based skill and knowledge base of graduating and practicing engineers.

The committee's membership was drawn from the corrosion engineering community, the materials engineering community, and engineering more broadly to provide the committee as a whole with a balanced opening perspective on the status and importance of corrosion education. All members of the committee (including those of us not from the corrosion field) were convinced by the evidence uncovered during the course of the study, first, that there is an urgent need to revitalize the corrosion education of our country's engineering workforce and, second, that revitalization can be accomplished only in the context of a healthy corrosion engineering research community.

The full committee met five times. First, the committee attended the NRC's Materials Forum: Corrosion Education in the 21st Century, a separately organized activity held in March 2007,[1] and met in closed session to consider how the views expressed at the forum impacted plans for the study. The committee's next three meetings had open sessions that featured discussions with the Office of Corrosion Policy and Oversight at DOD (at the first meeting) and with four invited panels of experts (two academic panels at the second and third meetings, one industrial panel at the third meeting, and one government agency panel at the second meeting). In response to the lack of consolidated data on corrosion courses in engineering curricula around the country, the committee circulated a questionnaire to university departments to elicit data that would allow it to assess the state of corrosion education. The committee gathered data on short courses, on publication trends, and on the makeup of the corrosion community. It also gathered opinions and information from representatives of the engineering and materials communities who made up the four panels mentioned above and at town meetings convened by the committee at the March 2008 meetings of The Minerals, Metals and Materials Society (TMS) and NACE International. While the panels and town meetings provided anecdotal information rather than hard data, the committee found the opinions and information compelling and consistent.

In addition, committee members gathered information between meetings with the assistance of two fellows of the National Academies' Christine Mirzayan Science and Technology Policy Graduate Fellowship Program: Vikas Khanna and Shikha Gupta.

The committee is grateful to all the participants at its plenary meetings: Reza Abbaschian, Aziz Asphahani, Kayle Boomer, Joseph Carpenter, Steven Carr, Robert Cottis, Dan Dunmire, Robert Hanrahan, William Hedges, Vincent Hock, Jo Howze, Bill Kepler, Alex King, Harriet Kung, Sue Louscher, Anthony Luscher, Michael J. Maloney, Robert Mroczkwski, Matthew O'Keefe, Michael R. Ostermiller, Michael Plumley, Vickie Plunkett, Ian Robertson, David Rose, Stephen Sharp, Robert Sinclair, Leslie Spain, Subra Suresh, Darrel Untereker, Peter Voorhees, Dan Walsh, and Ward Winer.

---

[1] See National Research Council, *Proceedings of the Materials Forum 2007: Corrosion Education for the 21st Century*, Washington, D.C.: The National Academies Press (2007). Available at http://books.nap.edu/catalog.php?record_id=11948. Accessed June 2008.

The committee became convinced early on of the importance of its task. It is widely believed that significant savings will accrue for this nation and that safety and preparedness will be improved if corrosion prevention and control are made a national priority and tackled as such. But doing so will require a workforce of engineers and technologists who are knowledgeable, savvy, and expert in corrosion science and engineering and its application. In addition, the committee became convinced that such a goal will only be achieved if there is significant buy-in from government, industry, and academia. This report lays out a plan for developing just that.

My personal thanks also go to the members of the committee for their considerable commitment of time and their efforts in writing this report. The committee is also grateful to Michael Moloney and Emily Ann Meyer of the NRC staff, who guided it through the study process.

Wesley L. Harris, *Chair*
Committee on Assessing Corrosion Education

Acknowledgment of Reviewers

This report has been reviewed in draft form by individuals chosen for their diverse perspectives and technical expertise, in accordance with procedures approved by the National Research Council's Report Review Committee. The purpose of this independent review is to provide candid and critical comments that will assist the institution in making its published report as sound as possible and to ensure that the report meets institutional standards for objectivity, evidence, and responsiveness to the study charge. The review comments and draft manuscript remain confidential to protect the integrity of the deliberative process. We wish to thank the following individuals for their review of this report:

Aziz Asphahani, Environmental Leverage, Inc.,
Rudy Bucheit, The Ohio State University,
Robert Cottis, University of Manchester,
L.S. (Skip) Fletcher, Texas A&M University,
Sylvia Hall, Ameron International,
Adam Powell, Oppenovation,
Mark Rhoads, GE Aviation,
Robert Sinclair, Stanford University,
Ponisseril Somasundaran, Columbia University,
Roger Staehle, Consultant, and
Ward Winer, Georgia Institute of Technology.

Although the reviewers listed above have provided many constructive comments and suggestions, they were not asked to endorse the conclusions or recommendations, nor did they see the final draft of the report before its release. The review of this report was overseen by Carl Lineberger, University of Colorado. Appointed by the NRC, he was responsible for making certain that an independent examination of this report was carried out in accordance with institutional procedures and that all review comments were carefully considered. Responsibility for the final content of this report rests entirely with the authoring committee and the institution.

Contents

SUMMARY   1

1   IMPORTANCE OF CORROSION ENGINEERING EDUCATION   9
   Understanding the Impact of Corrosion and Corrosion Engineering
      Education, 9
      Financial and Nonfinancial Costs of Corrosion, 10
      Why Corrosion Engineering Education Is Important for Designers,
        Purchasers, and End Users, 13
      Transportation Fuels Infrastructure, 15
      Engineered Devices and Systems, 17
      Energy Infrastructure, 17
      Health Care, 17
      Electronics and Computers, 18
      National Defense, 18
      Public Infrastructure, 19
      Historical Interest, 19
      In Our Homes, 20
      In Summary, 20
   Backdrop to the Study, 22
      Government Concern About Corrosion and Corrosion Engineering Edu-
        cation, 23
      Why a Corrosion Engineering Education Study Is Timely, 26

Role of Corrosion Research, 27
Scope of the Study—Metals and Nonmetals, 28
Outline of the Report, 30

2   AN ASSESSMENT OF CORROSION EDUCATION   31
Undergraduate Corrosion Education, 34
    The Dedicated Corrosion Course, 35
    Survey Course That Includes Corrosion, 38
    Senior Design Course, 39
    Discussion, 39
Graduate Corrosion Education, 48
Continuing Corrosion Education, 56
Summary of Findings, 60

3   CONCLUSIONS AND A RECOMMENDED PATH FORWARD   63
The Importance of Corrosion Education, 64
Consequences of the Current State of Corrosion Education, 65
    Graduating Engineers, 67
    Practicing Engineers in Government and Industry, 68
Recommendations for a Path Forward, 81
    Strategic Recommendations, 82
    Tactical Recommendations, 83

APPENDIXES

A   TWO EARLIER REPORTS   97
B   DATA GATHERED FROM UNIVERSITIES   106
C   PUBLICATIONS DATA   124
D   SHORT COURSES ON CORROSION   132
E   AGENDAS FOR MATERIALS FORUM 2007 AND COMMITTEE   146
    PUBLIC MEETINGS
F   SAMPLE LEARNING OUTCOMES   151
G   COMMITTEE BIOGRAPHIES   155

Summary

Corrosion has been the subject of scientific study for about 150 years. Historically, corrosion has meant the destructive oxidation of metals. But today engineering applications include a multitude of nonmetallic materials, and the term "corrosion" now signifies the degradation and loss of function by exposure to the operational environment of all materials. Corrosion can have a great impact on the safety and reliability of an extremely wide range of articles of commerce, and its economic impact in the United States is very large. It plays a critical role in determining the life-cycle performance, safety, and cost of engineered products and systems of value to the national defense and the general health and welfare of the public. Technology areas where corrosion plays an important role include energy production (for example, power plant operation and oil and gas exploration, production, and distribution), transportation (for example, automotive and aerospace applications), biomedical engineering (for example, implants), water distribution and sewerage, electronics (for example, chip wiring and magnetic storage), and nanotechnology. While the successful application of corrosion understanding already saves billions of dollars annually in these endeavors, studies have concluded that a wider application of our understanding of the corrosion phenomenon could reduce the cost of corrosion to the nation even more.[1]

---

[1]One such study, the Department of Defense's (DOD's) *Efforts to Reduce Corrosion on the Military Equipment and Infrastructure of the Department of Defense* (2008), estimates that the average return on investment from over 80 corrosion mitigation projects carried out over 3 years is around 50:1. Available at http://www.corrdefense.org/CorrDefense%20Magazine/Summer%202007/PDF/2007_DOD_Corrosion_Report.pdf. Accessed August 2008.

The 2001 report *Corrosion Costs and Preventive Strategies in the United States* noted that technological changes and the wider use of available corrosion management techniques have improved corrosion mitigation.[2] However, better corrosion management can also be achieved using preventive strategies in nontechnical and technical areas. These preventive strategies include (1) increase awareness of the large costs of corrosion and the potential for savings, (2) change the misconception that nothing can be done about corrosion, (3) change policies, regulations, standards, and management practices to decrease corrosion costs through sound corrosion management, (4) improve education and training of staff in recognition and control of corrosion, (5) improve design practices for better corrosion management, (6) advance life prediction and performance assessment methods, and (7) advance corrosion technology through research, development, and implementation. Although there are likely to be many reasons why these strategies are not routinely followed, in the committee's view strengthening corrosion education would be a major step toward improved corrosion control and management. An engineering workforce that is ill-equipped to deal with corrosion problems begs the question, What are engineers being taught about corrosion? Is it sufficient? This study was commissioned to do two things:

- Assess the level and effectiveness of existing engineering curricula in corrosion science and technology, including corrosion prevention and control, and
- Recommend actions that could enhance the corrosion-based skill and knowledge base of graduating and practicing engineers.

From the perspective of assessing corrosion education, the workforce of graduating and practicing engineers is divided as follows:

- Technologists who perform repeated critical tasks;
- Undergraduate engineering students in materials science and engineering (MSE), who upon graduation should be knowledgeable in materials selection;
- Undergraduate students in other engineering disciplines; and
- MSE graduate students, who upon graduation should be very knowledgeable in materials selection.

Advances in corrosion control are integral to the development of technologies that can solve the engineering grand challenges related to the sustainability

---

[2]For more information, see www.corrosioncost.com/home.html. Accessed February 2008.

and vulnerability of current, legacy, and future engineered products, systems, and infrastructure. Some examples are as follows:

- *Energy infrastructure.* Corrosion is likely to be a key issue in solar cell lifetime and wind turbine performance. Strategies for scrubbing emissions and capturing carbon will likely be limited by corrosion, as will be the high efficiencies of central power plants, which are achieved by means of a high-temperature working fluid.
- *Transportation fuels infrastructure.* A new set of corrosion problems is likely to limit the operation of the infrastructure for the production and delivery of new fuel/power systems that rely on batteries, fuel cells, hydrogen, ethanol-based biofuels, and so on.
- *Engineered devices and systems.* Composite structures, ceramics, and reactive metals (such as magnesium) require better corrosion protection because they are less tolerant of corrosion. For instance, lightweight magnesium, a key technology being developed by the auto industry, is considerably more reactive than steel or aluminum. Graphite composites require greater environmental resistance to maintain structural integrity.
- *Health care.* The drive to minimize size, maximize capability, and extend medical device lifetimes places demands on the materials of construction and on their tolerance for degradation before function is affected. As more medical devices are implanted to protect and assist an ever-aging population, unexpected uses and failures continually occur, and improved understanding of the durability of such implanted devices will depend on their designers having extensive training in corrosion science.
- *Electronics and computers.* As modern electronic circuitry goes to ever smaller dimensions, new problems arise from environmental attack on circuits as their surface to volume ratio increases. Sensors are of growing importance in daily life—from monitoring biological activity in the body, to controlling our cars and providing information on environmental conditions such as wind, precipitation, and chemical contamination. Surface and interface corrosion processes in these sensors will therefore pose an increasing threat to device and system reliability.
- *National defense.* Defense readiness is highly sensitive to corrosion, and future defense systems will continue to present fresh challenges as new materials are inserted into defense platforms. At any given time, 20 to 50 percent of the U.S. Air Force tanker fleet is in repair; many U.S. Army vehicles are in repair or are being used at less than full capacities owing to general wear and corrosion.

In general, materials being used in the modern world are being pushed to the limits of their operability. The demands will require a workforce conscious of environmental attack on all types of systems and having the ability to anticipate and design for sustainability under extreme conditions. The engineering workforce must have a solid understanding of the physical and chemical bases of corrosion, as well as an understanding of the engineering issues surrounding corrosion and corrosion abatement.

The study revealed, nonetheless, that corrosion engineering is not a required course in the curriculum of most bachelor's degree programs in MSE and related engineering fields. In many programs, corrosion is not only not a required subject, it is not even available. As a result, most bachelor's-level graduates of materials- and design-related programs have an inadequate background in corrosion engineering principles and practices.

Employers recognize the need for employees with competence in corrosion engineering, but they are not finding it in today's graduates. Indeed, their principal concern is that those making design decisions "don't know what they don't know" about corrosion. In the committee's judgment, this lack of knowledge and awareness ultimately jeopardizes the health, wealth, and security of our country.

This report also reminds us of the obvious: that the availability of corrosion classes for graduating and practicing engineers depends on the availability of people to teach the subject. The availability of teachers is in turn dependent on the health of the corrosion research community and therefore on the research support available to that community. If corrosion engineering education is to flourish, the committee believes the number of MSE faculty specializing in corrosion will need to increase. This means that federal agencies and industry will need to support university-based corrosion specialists, who will become a foundational corps of teachers.

The committee has found that industry compensates for the inadequate corrosion engineering education of practicing engineers through on-the-job training and short courses for its employees and the hiring of outside consultants as required. These continuing skills-based and knowledge-based educational approaches are widely accepted as useful, and they play an important role depending on the job function and desired outcomes. However, the continuing education of the workforce is not a substitute for including corrosion in the curricula for graduating engineers and technologists.

In government agencies such as DOD, the Army Corps of Engineers, the Department of Energy, the U.S. Bureau of Reclamation, and state transportation agencies, the committee finds that maintaining a corps of in-house corrosion experts is not now and has probably never been a high priority. Likewise, the committee's sense is that current management philosophy in government appears to expect project managers to find a corrosion expert on demand when projects require that expertise, largely by outsourcing to a contractor or consultant.

Industry and government reliance on outside contractors to conduct the continuing education of the workforce or to act as corrosion consultants is ultimately unsustainable as these outsiders learned their trade in the industries and agencies that are now buying in their services and that are no longer employing (and hence training) their successors. This situation is aggravated by the retirement of the few people with corrosion expertise and the near absence of corrosion engineering experience in new hires emerging from graduate and undergraduate engineering programs.

Based on the committee's expert judgment and its assessment of the data gathered during the course of this study and the opinions and information received from government, industry, and the MSE and engineering communities, the committee concludes that the current level and effectiveness of engineering curricula in corrosion, offered through university-based and on-the-job training, will not provide a sufficient framework to allow the country to reduce substantially the national cost of corrosion or to increase the safety and reliability of the national infrastructure. In addition, the committee concludes that the recent proactive stance on corrosion control that DOD has taken will be undermined by the shortage of engineers and technologists with a sufficient comprehension of corrosion. To enhance the corrosion-based skill and knowledge base of graduating and practicing engineers, the committee concludes that corrosion education needs to be revitalized through (1) short-term tactical actions by educators, government, industry, and the broader technical community and (2) long-term strategic actions by the federal government and the corrosion research community. The committee is not recommending a wholesale overhaul of engineering education. Rather, it has identified a series of actions that can be adopted by institutions—educational, governmental, and community—that are interested in increasing corrosion education and awareness. While acknowledging that there are many pressures on the curricula in the country's engineering schools, the committee hopes many universities' departments of engineering and MSE will acknowledge the importance to the country of improving the provision of corrosion knowledge to our future engineers.

THREE SHORTER-TERM TACTICAL RECOMMENDATIONS

Recommendation: Industry and government agencies should strengthen the provision of corrosion engineering education. They should

- **Develop a foundational corps of corrosion faculty by supporting research and development in the field of corrosion science and engineering.**
- **Provide incentives to the universities, such as endowed chairs in corrosion control, to promote the hiring of corrosion experts at the universities.**
- **Enable the setting and periodic updating of learning outcomes for corrosion courses by publishing and publicizing skills sets for corrosion technologists and engineers.**

- Fund the development of educational modules for corrosion courses.
- Support faculty development, offering corrosion-related internships and sabbatical opportunities, and supporting cooperative programs between universities and government laboratories to facilitate the graduate student research experience.
- Increase support for the participation of their engineers in short courses when specific skills shortages are identified and are required to be filled in the short term.

Recommendation: Engineering departments in universities should incorporate elective learning outcomes and course work on corrosion into all engineering curricula. Improving the overall awareness of corrosion control will require that more engineers have a basic exposure to corrosion, enough to "know what they don't know."

Recommendation: Materials science and engineering (MSE) departments in the universities should introduce set required learning outcomes on corrosion into their curricula. All MSE undergraduate students should be required to take a course in corrosion control so as to improve the corrosion knowledge of graduating materials engineers.

TWO LONGER-TERM STRATEGIC RECOMMENDATIONS

In addition to the recommendations above, the details of which are expanded on in the report, during the course of the study the committee became convinced that there were two compelling challenges outstanding: one for the federal government, in particular the DOD, and one for the corrosion community itself.

The committee is convinced that government can improve the education of the corrosion workforce by developing a strategic plan with a well-defined vision and mission. This is the first long-term recommendation the committee is making, and it is directed to DOD, specifically its Corrosion Policy and Oversight Office. This plan will require input from a broad set of societal stakeholders and the analytical capabilities of government, industry, and academia.

Strategic Recommendation to the Government

Recommendation: The Department of Defense's Director of Corrosion Policy and Oversight, whose congressionally mandated role is to interact directly with the corrosion prevention industry, trade associations, other government corrosion-prevention agencies, academic research and educational institutions, and scientific organizations engaged in corrosion prevention, should (1) set up a corrosion education and research council composed of government agencies, industry, and academia to develop a continuing strategic plan for fostering corrosion education and (2) should identify resources for executing the plan. The plan should have the following vision and mission:

- *Vision.* A knowledge of the environmental degradation of all materials is integrated into the education of engineers.
- *Mission.* To provide guidance and resources that will enable educational establishments to achieve the vision.

The challenge to the corrosion community is motivated by the committee's observation that the community appears isolated from the rest of the scientific and engineering community. Repeatedly the committee heard that, on the one hand, the general research and engineering community considers corrosion science and engineering to be a mature field, with little compelling science to be done, and that on the other hand, the corrosion community considers there are many compelling science questions to be answered, with corrosion mitigation and prevention likely to be considerably advanced if these questions can be answered.[3] The responsibility for changing this mismatch in perception falls to the corrosion community itself. Because the education of a corrosion-savvy workforce is dependent, broadly, on the health of the corrosion community, the committee offers its second long-term recommendation to this community.

Strategic Recommendation to the Corrosion Community

Recommendation: To build an understanding of the continuing need for corrosion engineering education, the corrosion research community should engage the larger science and engineering community and communicate the challenges and accomplishments of the field. To achieve this goal the corrosion research

---

[3] A National Research Council study getting underway in the autumn of 2008 is charged with identifying the most compelling scientific questions in fundamental corrosion science. The kinds of questions it will be considering include these: Is enough known about corrosion to enable the lifetime of a material to be increased by a factor of 5 or 10? What is the mechanism of pit initiation? What are the next important processes in corrosion to understand better and model? What is the true chemistry inside localized corrosion sites, and how does it affect the corrosion processes? Corrosion at the nanoscale: What is really of interest, and can corrosion at the nanoscale be forecast from first principles and multiscale knowledge? What fundamental theory or model toolkit capability do we need to develop? For more information, see http://www.nationalacademies.org/nmab. Accessed August 2008.

community should identify and publish the research opportunities and priorities in corrosion research and link them to engineering grand challenges faced by the nation. To show how the field of corrosion could meet these challenges, the corrosion research community should reach out to its peers by speaking at conferences outside the field, publishing in a broad range of journals, and writing review articles for broad dissemination.

1

Importance of Corrosion Engineering Education

This is a report on corrosion engineering education. The field of corrosion is concerned with the change over time of all engineering solids, including metals, ceramics, glasses, polymers, aggregates, composites, and other materials as they are exposed to environments such as chemicals, stress, radiation, and so on. It is impossible to consider the importance of corrosion engineering education without also looking from a broad perspective at the impact corrosion has on the United States. This chapter, as an introduction to this report, will discuss how that impact motivates concern about the education of the workforce that is either battling corrosion or should be.[1] That concern is what led to this report being commissioned.

UNDERSTANDING THE IMPACT OF CORROSION AND CORROSION ENGINEERING EDUCATION

The continued reliability and safety of the U.S. industrial complex and public infrastructure are essential to the nation's quality of life, industrial productivity, economic competitiveness, and security and defense. Each component of the public infrastructure—highways, airports, water supply, waste treatment, energy supply, power generation, etc.—is part of a complex system requiring significant investment. Within that infrastructure, in both the private and government sectors, corrosion affects nearly all of the materials and structures used. Corrosion, therefore,

---

[1]Note that this report focuses on postsecondary education—that is, at the university, college, and workforce continuing education levels.

affects us in everyday life—in the manufacture of products, the transportation of people and goods, the provision of energy, the protection of our health and safety, and the defense of the nation. By discussing the impact of corrosion, this section sheds light on the importance of teaching engineers about corrosion. Figure 1-1 shows a single but vital element of the national infrastructure, an offshore semi-submersible drilling rig that is undergoing intense corrosion as a result of its exposure to saltwater and the moisture-laden, chloride-containing atmosphere.

Financial and Nonfinancial Costs of Corrosion

Although most people think of rust when they think of corrosion, the term refers not only to the oxidization of iron but can also refer to the degradation of all materials (metals, polymers, ceramics, semiconductors, and so on) that make up the public infrastructure and physical systems as diverse as the nation's highway network, its military equipment, and medical devices implanted in our bodies.[2] The costs associated with corrosion, although largely hidden, are borne by every consumer, user, and producer. They are enormous, estimated to be 3.1 percent of the U.S. gross domestic product (GDP).[3] Applying this percentage estimate to the 2007 GDP of about $14 trillion gives a cost in 2007 dollars of $429 billion.[4] With a population of 303 million, that works out to $1,416 per person per year in the United States. This estimate is supported by similar estimates in other countries (Box 1-1). The effects of corrosion on safety, health, and the environment are not so readily quantifiable, but failures of infrastructure illustrate the potential for severe impacts on daily life and the economic health and security of the nation.

The importance of mitigating corrosion is not just about saving money. It is equally—and in some cases more importantly—about readiness. Operating equipment in severe or unexpected environments can exacerbate corrosion and make systems unreliable. Readiness is critical in such systems as military equipment, emergency response systems, or very specialized systems like the launch facilities of the National Aeronautics and Space Administration (NASA). Often problems

---

[2] For the purposes of this report, the committee will use the term corrosion to refer to the deterioration of a material in its operating or usage environment. NACE International, known as the National Association of Corrosion Engineers when it was established in 1943 by 11 corrosion engineers in the pipeline industry, defines corrosion as the deterioration of a material, usually a metal, that results from a reaction with its environment. While corrosion is associated mostly with metals, the committee considers corrosion to include the degradation of all materials—including polymer, composite, and ceramic materials—that results from a reaction with the environment.

[3] For more information, see the Federal Highway Administration study *Corrosion Costs and Preventive Strategies in the United States* (1999), hereinafter called *Corrosion Costs.* Available at www.corrosioncost.com/pdf/main.pdf. Accessed February 2008. The report is summarized in Appendix A.

[4] GDP data from http://www.bea.gov/national/index.htm#gdp. Accessed April 2008.

FIGURE 1-1 Offshore semisubmersible drilling rig used in the production of oil and gas. Courtesy of Richard Griffin.

occur in locations not easily visible to system operators or users. Figure 1-2 exemplifies this by showing corrosion under insulation in the bilge area of a naval ship. Reliability is also of critical importance in health- and safety-related applications such as biomedical implants, electronics, and sensors.

Those organizations, both governmental and private, producing, operating, or maintaining physical structures, objects, and the built environment know that they should be designing systems with protection against environmental degradation. The true cost of any system that has a lifetime longer than a modern commoditized product like a cell phone must take into account corrosion protection, maintenance, and system performance monitoring.

In addition to the dollar impact of corrosion there are impacts that are difficult to quantify in terms of money. These are the effects of corrosion on the environment and on society. Clearly, leakage of chemicals, oil, or sewage from corroded tanks, drums, and pipes can have long-term effects on the environment, including the water supply, air quality, contamination of food crops, livestock, buildings, and the wildlife population. Slow leaks of underground fuel tanks have been an ongoing problem; the cost of replacement and decontamination can be calculated

BOX 1-1
Studies on the Cost of Corrosion

A number of studies have been done in the past 60 years in an effort to quantify the losses attributable to corrosion (Table 1-1-1). In 1950, Uhlig estimated that the cost to the United States was about 2.1 percent of the GDP. Similarly, the Hoar report in the United Kingdom showed that corrosion cost amounted to about 3.5 percent of the GDP. The ramifications of this study resulted in the creation of the University of Manchester Institute of Science and Technology Corrosion Centre. A 1974 study for Japan, on the other hand, showed the cost of corrosion was 1.2 percent of its gross national product (GNP). In 1975 for the United States, the National Bureau of Standards and the Battelle Research Institute concluded that the cost of corrosion was 4.5 percent of the GNP. The most recent study, by the U.S. Federal Highway Administration (FHWA) and completed in 2002, estimated the cost of corrosion in 1998 to be 3.1 percent of the U.S. GDP—that is, $276 billion. (This study, referred to in this report as *Corrosion Costs*, is summarized in Appendix A.) Even though there is scatter in these numbers and it is likely that the 2002 GNP is much more heavily weighted to the service economy and less heavily to materials, assets, and maintenance costs than the 1975 GNP, there can be no denying that the impact of corrosion and environmental degradation on the economies of the developed nations is considerable.

TABLE 1-1-1 Studies on Cost of Corrosion

Year of Publication	Author	Country	Share of Country's GDP (%)
1950	H.H. Uhlig	United States	2.1
1970	T.P. Hoar	United Kingdom	3.5
1974		Japan	1.2[a]
1975	Battelle/NBS	United States	4.5[a]
2000	DTI	United Kingdom	2.5-3.5
2002	NACE/FHWA	United States	3.1

[a]Share of GNP.

in dollars, but the long-term effects of contamination of large areas of land cannot be represented simply in dollars. Leaking underground storage tanks are a source of pollutants at many Superfund sites—sites whose cleanup is time consuming and expensive and restricts the use of land and water for many years.

In addition to the effects on human and wildlife health and survival of leaks resulting from corrosion, there is the societal cost of injury or death from corrosion-related accidents such as bridge collapses or the corrosion-related failure of medical implants and devices. Whether or not such effects are quantifiable, the overall loss to society and the environment must be considered when assessing the impact of corrosion.

FIGURE 1-2 Corrosion under insulation in the bilge area of a naval ship.

While humanity's footprint on the global environment may need to be minimized, there will always be a built environment that needs to be protected, improved, and maintained for the safety and well-being of all who live in it. Networks and systems for power transmission, water delivery, and information flow are the fabric that holds modern civilization together, and they call for reliability, maintainability, and sustainability. That corrosion is jeopardizing our nation's economy, defense, health, and environment is now well documented and motivates a closer look at what this country's engineers, technicians, and other practitioners who design, manufacture, build, and maintain the national infrastructure are learning about corrosion.

Why Corrosion Engineering Education Is Important for Designers, Purchasers, and End Users

Designers are responsible for creating products that can perform safely, reliably, and efficiently. They must understand and take into account the operating condi-

FIGURE 1-3 Corrosion on an automobile operating in a warm, moist chloride environment. While anticorrosion technology for cars has improved tremendously in recent years, this picture shows the extreme corrosion that is still possible. Courtesy of Richard Griffin.

tions and possible failure modes. Not anticipating and mitigating corrosion can expose products to a high risk of failure. In the auto industry, severe corrosion of the auto body was once a considerable problem but has been largely overcome by using corrosion-resistant materials and corrosion prevention strategies. Figure 1-3 shows what can happen to an automobile in an aggressive environment.

Degradation of structures can result in loss of production or usefulness and, in some cases, of life. The consequences can be very costly and include the possibility of product liability suits. To mitigate these risks, a knowledgeable assessment of the causes and possible prevention of corrosion-related failure or degradation should be incorporated into the design early on. In current design practice, this assessment usually takes the form of a failure modes and effects analysis (FMEA). For corrosion to be adequately addressed, designers need to know about the mechanisms of corrosion failure and to know when they need assistance from a corrosion specialist in selecting materials or operating parameters.

Purchasers have the responsibility for ensuring that their chosen system functions safely and efficiently, and they bear the financial burden of its maintenance.

Those responsible for maintaining our public infrastructure need a sufficient store of knowledge about corrosion engineering to recognize that corrosion has been taken into account in the design process and to estimate the costs of operating the product they have purchased, including the development of a realistic and proactive plan for corrosion mitigation. Roads, bridges, planes, pipeline systems, and the electrical grid are examples of systems where a comprehensive understanding of corrosion can lead directly to lower maintenance costs, longer service lifetimes, and less risk of failure.

End users may have little direct connection to the designer but must be confident that the designer has taken corrosion into account. They—and the public at large—must implicitly trust that designers and suppliers are providing goods that are useful for the stated purpose and safe as well. While the monetary cost of corrosion can be estimated, the cost of risks to public safety cannot be so easily measured without performing a complex risk assessment. Public safety and the environment are the main reason end users and the public should be concerned about the state of corrosion engineering education and the implementation of that education by the engineering design workforce.

From the evidence the committee examined, corrosion will clearly continue to have a major impact on key industries and infrastructure systems being planned. How industries function and how systems are built will be strongly influenced by their response to the environment in which they must operate. As discussed above, designers of devices and structures, those who purchase them or maintain them, and, of course, their end users should at the very least be aware of the impact of corrosion. The committee hopes that these stakeholders will be among the readers of this report and draws their attention to the following discussion about why they should be interested in corrosion engineering education. The sections that follow describe the crucial role of corrosion in the infrastructure systems that are the lifeblood of the U.S. economy.

Transportation Fuels Infrastructure

The overdependence on oil of the nation's transportation system and the pace of global climate change will bring about radical transformations in energy supply and use in the years ahead. The next decade may see the gradual electrification of the automobile with its concomitant dependency on fuel cells, batteries, and the electricity grid. A new set of corrosion problems accompanying this transformation is likely to limit the service lives of the batteries and fuel cells. The production of hydrogen in quantities large enough to make an impact on the transportation infrastructure will probably require thermochemical or electrochemical processes using very high temperatures and highly corrosive solutions. The storage tanks for hydrogen will have to resist degradation such as hydrogen embrittlement.

FIGURE 1-4 An internal pit in an oil field water injection pipe. Courtesy of Richard Griffin.

The storage tanks and pipeline systems for internal combustion engines that are designed to run on ethanol-based biofuels will need to be redesigned, and gasoline additives will need to be developed because of unique corrosion issues and the affinity of these biofuels for water. The present pipeline infrastructure for delivering natural gas, crude oil, and refined gasoline products is also operating at its limits. For example, there have recently been significant failures associated with corrosion at Prudhoe Bay, Alaska.[5] Figure 1-4 shows an internal corrosion pit that was found in a water injection system in an oil field. In addition, modifications to gasoline formulations that use renewable resources (ethanol) have now made these formulations more sensitive to water uptake and increased the potential for corrosion during storage and transport.

---

[5] For more information, see http://www.bp.com/genericarticle.do?categoryId=2012968&contentId=7020563. Accessed March 2008. Also see http://www.petroleumnews.com/pntruncate/573947058.shtml. Accessed March 2008.

Engineered Devices and Systems

Many new engineering structures employ lightweight materials to save money and energy. Composite structures, ceramics, and reactive metals (such as magnesium) result in less mass in the finished product, but they might require better corrosion protection. Lightweight magnesium is considerably more reactive than steel or aluminum. Graphite composites can be made into very thin, stiff, lightweight structures but will require greater environmental resistance to maintain structural integrity. The push for extended component lifetimes and less design effort means that environmental degradation will become a greater concern.

Energy Infrastructure

The nation and the world will be challenged to rebuild the energy infrastructure in a way that avoids greenhouse gas emission and maximizes efficiency of electricity production. Wind turbines, solar cells, biofuels, nuclear energy, and clean coal are all set for significant development and increased implementation. Corrosion is likely to be a key issue in solar cell lifetime and wind turbine performance and will become more important in large central power plants. Strategies for scrubbing emissions and capturing carbon will likely be limited by corrosion. Similarly, high efficiencies of central power plants are achieved through the use of very high temperature working fluids, which means much more expenditure for corrosion protection than we see today. As the country considers commissioning more nuclear power plants, the storage and (eventual) disposal of nuclear waste are largely an issue of containment vessel corrosion rates and possible failure modes. Virtually all energy sources will see an increasing cost of corrosion and new forms of corrosion. Therefore, developments in corrosion technology are key to improved efficiency in energy production.

Health Care

Health care is increasingly dependent on biomedical devices that monitor and control bodily functions and deliver drugs. The drive to minimize size, maximize capability, and extend device lifetime places demands on the materials of construction and on their tolerance for degradation before function is affected. The use of new materials over longer and longer times will require knowing more about the interaction between these materials and the human body environment. Their exposure to drugs will call for such devices to have particularly high resistance to chemical interactions. New uses of such devices and implants are limited by the need for them to resist corrosion for extended lifetimes. As more medical devices are implanted to serve an ever-aging population, unexpected uses and failures can

FIGURE 1-5 Corroded circuit board. Courtesy of Richard Griffin.

occur. However, designers will not be able to solve these problems unless they have extensive training in corrosion science.

Electronics and Computers

As modern electronic circuitry is reduced to ever-smaller dimensions in order to increase memory and computational density, new problems arise from environmental attack on circuits as their surface to volume ratio increases. The insertion of smart devices and the more widespread the deployment of sensors in all types of systems and structures puts electronics into ever-harsher use environments. Although a cell phone is now a commodity product, the owner still expects that the instrument will reliably switch micro-amp currents through its various switches for a very long time. Corrosion on these contacts can destroy instrument quality and reliability. Sensors are ever more important in daily life, from monitoring biological activity in the body to controlling our cars and providing information on wind, precipitation, chemical contamination, and so on. The increase in sensor utility is driven by advances that shrink the devices to the micrometer level or less, resulting in significantly larger surface areas for the same active volume. The result is that surface and interface corrosion processes will become much more important than they are today and will pose an increasing threat to device and system reliability. Figure 1-5 shows corrosion of a circuit board that had been in a data logger used in a moist environment.

National Defense

Defense readiness is highly sensitive to corrosion, and future defense systems will still present new challenges as new materials are inserted into defense platforms. Ground vehicles designed for cold war battlefields are being used in desert environ-

FIGURE 1-6 Examples of corrosion on various bridges. *Left:* corrosion on the historic Devil's Elbow Bridge on Missouri Route 66. Courtsey of Conor Watkins, Missouri University of Science and Technology. Available at http://www.rollanet.org/~conorw/cwome/article51&52combined.htm. *Right:* Rebar corrosion, bridge on 401 Highway in Ontario, Canada. Courtesy of Tim Mullin. Available at www.corrosion-club.com/rebarimages.htm.

ments, where the degradation modes are different. The use of smart materials on the battlefield will require robust resistance in aggressive environments. The future soldier will probably be clad in multifunctional uniforms that possess communication capability, power sources, and armor, all of which will require considerable innovation in durability. Exceedingly important from the standpoint of national defense is the impact that corrosion damage has on overall defense readiness. At any given time, 20 to 50 percent of the U.S. Air Force tanker fleet is in repair; many U.S. Army trucks and HMMWVs are in repair or are being used at less than full capacities owing to general wear and corrosion.[6] Superimpose on this the huge amount of delayed maintenance and repair of weapon systems and infrastructure of the Department of Defense (DOD) due to the Iraq conflict, and one can see clearly that corrosion, wear, and general systems degradation represent a significant cost for DOD, in the tens of billions of dollars annually. Little is being done to train and prepare present and future professionals in handling this problem properly, and DOD struggles to train its workers to deal well with corrosion.

Public Infrastructure

The national infrastructure and its maintenance is an important issue (see Figure 1-6). The Bureau of Reclamation at the Department of the Interior faces

---

[6]See the following reports from the Government Accountability Office (GAO) at http://www.gao.gov/cgi-bin/getrpt?GAO-06-709 and the Defense Science Board at http://handle.dtic.mil/100.2/ADA428767. Accessed March 2008.

serious corrosion-related issues in handling water storage and transport in the Western states, and all municipal water treatment and delivery agencies and infrastructure systems face significant water losses up to the point of delivery due to pipes damaged by corrosion. The corrosion of bridges, decks, and the steel-reinforced concrete structures of our highway system endangers public safety and incurs large outlays every year for such maintenance, which could be reduced by paying more attention to corrosion control at the design stage. The sustainability of the modern infrastructure depends on the proper design and maintenance of its major components, and this in turn demands a cadre of engineers and scientists capable of choosing and using materials so as to minimize environmental degradation.

Historical Interest

Even our historical artifacts are constantly undergoing degradation and must be maintained to preserve them for present and future generations. Structures and objects from the past require special handling and restoration and protection methods. Bronze statuary of historical interest has become a special problem recently as acid rain has begun to attack bronze alloys that were formerly inert to environmental attack (Figure 1-7). The Statue of Liberty was found to have suffered considerable corrosion when it was restored in the early 1980s.[7] Each of the 1,350 shaped iron ribs backing the statue's skin had to be removed and replaced when it was discovered that the iron had experienced galvanic corrosion wherever it contacted the copper skin, losing up to 50 percent of its thickness.

In Our Homes

Within our homes, as new building materials are introduced, different corrosion hazards present themselves. Composites that do not degrade like natural materials or metals will be used increasingly in building construction. Connecting the copper piping of a home to steel mains is difficult, as is the incorporation of magnesium anodes for protection of water heaters. See Figure 1-8 for an example of a corroded water heater.

In Summary

In general, we are pushing the limits of operability with all of the materials we use in the modern world. For instance, the United States is committed to putting manned communities on the moon and Mars and must develop transportation

---

[7] R. Baboian, E.L. Ballante, and E.B. Cliver, *The Statue of Liberty Restoration*, Houston, Tex.: NACE International (1990).

FIGURE 1-7 Corrosion on Rodin's "Thinker." Courtesy of Philadelphia Museum of Art: Bequest of Jules E. Mastbaum, 1929.

and support systems to meet the challenges involved, including those posed by some very unsparing radiation environments. Humanity is expanding into harsher environments on Earth, requiring systems, objects, and structures that can support human activity at great ocean depths, deep underground, and in desert and arctic environments. All of these demands will require a workforce conscious of environmental attack on all types of systems and able to anticipate and design for sustainability under extreme conditions.

One of the biggest driving factors is the trend to extend the useful lifetimes of items beyond their original design lifetimes. It is rare to stop using a bridge, for example. Commercial and military aircraft in daily use in the United States are operating well beyond their expected lifetime (Figure 1-9). With a bridge, about 10 percent of the construction cost of the structure controls the lifetime cost. The components of lifetime expense are the costs associated with replacement, readiness, safety, and reliability. The choice often boils down to "pay me now or pay me more later" in design, materials choice, and maintenance. The key to bringing that perspective to design and manufacturing is educating the nation's engineers.

FIGURE 1-8 Corroded residential hot water heater. Courtesy of Richard Griffin.

BACKDROP TO THE STUDY

Proactive corrosion prevention and control can lead to savings in the cost of dealing with corrosion in every area of the economy. But all too often such practices are not employed because of a shortfall in investments or a lack of knowledge on the part of designers. While better and more cost-effective corrosion management procedures could significantly extend the service life of existing systems and reduce maintenance costs and replacement requirements, the value of preventive strategies is often not recognized and they are not even applied. The widespread misconception that nothing can be done about corrosion is exacerbated by the fact that no one is "selling" corrosion. There is no identifiable advocate for corrosion control as there is for, say, the steel or aluminum industries. While there are interested parties, such as the corrosion mitigation industry and professional societies like NACE International, corrosion is not a product per se and there is no national advocate for corrosion programs.

FIGURE 1-9 On April 28, 1988, Aloha Flight 243 took off from Hilo, Hawaii, bound for Honolulu. As it reached its flight altitude, the cockpit crew heard a loud noise and looked back to see that a part of the passenger roof had blown off the aircraft. The skilled crew landed the plane safely with only one death. The aircraft had been designed to sustain major structural failure and survive. The National Transportation Safety Board (http://www.ntsb.gov/publictn/1989/AAR8903.htm) determined that the probable cause of this accident was the failure of the Aloha Airlines maintenance program to detect the significant disbonding and fatigue damage, which ultimately led to failure of the lap joint and the separation of the fuselage's upper lobe. Also contributing to the accident were the failure of the Federal Aviation Administration (FAA) to require inspection of all the lap joints proposed by Boeing after the discovery of early production difficulties in the 737 cold bond lap joint, which resulted in poor bond durability, corrosion, and premature fatigue cracking.

Government Concern About Corrosion and Corrosion Engineering Education

In government circles there is a growing recognition of the need for a better understanding of corrosion and its mitigation. The federal government is investing more resources to preserving the nation's infrastructure, security, and defense systems and to understanding the true cost of their maintenance. Figure 1-10 illustrates damage done to the flight deck of an aircraft carrier, an example of what corrosion can do to DOD assets.

DOD expresses its interest in corrosion as follows:[8]

---

[8]See http://www.corrdefense.org/CorrDefense%20WebPage%20Content/WhyDODMustProtectItsAssets.aspx. Accessed February 2008.

FIGURE 1-10 Corrosion on the flight deck of a naval aircraft carrier.

The Department of Defense acquires, operates, and maintains a vast array of physical assets, ranging from aircraft, ground vehicles, ships, and other materiel to buildings, airfields, ports, and other infrastructure. Furthermore, in order to perform its mission, DOD must train and fight in all environments, including some of the most corrosively aggressive on Earth. Consequently, DOD assets are subject to significant degradation due to corrosion, with specific effects in the following areas:

- Safety—A number of weapon system mishaps have been attributed to the effects of corrosion. For example, corroded electrical contacts on F-16s caused "uncommanded" fuel valve closures (with subsequent loss of aircraft), and corrosion-related cracking of F/A-18 landing gears resulted in failures during carrier operations.
- Readiness—Weapon systems are routinely out of commission due to corrosion deficiencies. For example, corrosion has been identified as the reason for more than 50 percent of the maintenance needed on KC-135 aircraft.
- Financial—The cost of corrosion to the DOD alone is estimated to be between $10 billion and $20 billion annually.

For these reasons, DOD has a long history of corrosion prevention and control. The Department has been a leader in many areas of research (ranging from understanding the fundamentals of corrosion to applying advanced materials, coatings, inhibitors, and cathodic protection for corrosion control).

DOD is taking action in this area now. It is, for instance, making its concerns felt in the area of military equipment and systems acquisition. A new procurement instruction, DODI 5000.2 (Operation of the Defense Acquisition System), addresses the impacts of corrosion for all procurements over $1 million.[9] These developments are linked to the new focus on corrosion found in the FY2007 DOD Authorization Act, which created the position of a corrosion executive and established a policy and oversight office at the Pentagon.[10] The legislation instructed that office to draw up a strategy on corrosion prevention and mitigation to include the following:

(A) Expansion of the emphasis on corrosion prevention and mitigation within the Department of Defense to include coverage of infrastructure.
(B) Application uniformly throughout the Department of Defense of requirements and criteria for the testing and certification of new corrosion-prevention technologies for equipment and infrastructure with similar characteristics, similar missions, or similar operating environments.
(C) Implementation of programs, including supporting databases, to ensure that a focused and coordinated approach is taken throughout the Department of Defense to collect, review, validate, and distribute information on proven methods and products that are relevant to the prevention of corrosion of military equipment and infrastructure.
(D) Establishment of a coordinated research and development program for the prevention and mitigation of corrosion for new and existing military equipment and infrastructure that includes a plan to transition new corrosion prevention technologies into operational systems.

Also relevant is the Senate Armed Services Committee report[11] directing the Secretary of Defense, working through the DOD corrosion executive and its Corrosion Policy and Oversight Office, to commission a study under the auspices of the National Research Council (NRC) to assess corrosion engineering education in engineering programs and develop recommendations that could enhance corrosion-related skills and knowledge, which, of course, is the mandate for the present study. In response, the NRC organized the Committee on Assessing Corrosion Education, which was charged to assess the level and effectiveness of existing engineering curricula in corrosion science and technology, including corrosion

---

[9] For a copy of the instruction, visit http://www.corrdefense.org/Key%20Documents/DODI%205000-2.pdf. Accessed February 2008.

[10] 10 U.S. Code 2228. Available at http://www.corrdefense.org/Key%20Documents/10%20U.S.C.%202228. pdf. Accessed February 2008.

[11] Available at http://thomas.loc.gov/cgi-bin/cpquery/T?&report=sr254&dbname=109&. Accessed August 2008.

prevention and control, and to recommend actions that could enhance the corrosion skills and knowledge base of graduating and practicing engineers.

Why a Corrosion Engineering Education Study Is Timely

Issues in corrosion engineering are similar but not identical to those encountered in the casting or steelmaking processes. As discussed earlier in this chapter, corrosion subtracts value from materials, while casting and steelmaking add value. All materials degrade with time in their environments. A basic understanding of this process is crucial to the education of the nation's scientists and engineers. That necessity notwithstanding, a distinction must be made between engineers who should be knowledgeable in materials selection generally and in corrosion specifically (that is, materials engineers) and other engineers who should be aware of corrosion issues but need not be particularly knowledgeable or expert (that is, chemical, mechanical, and civil engineers).

As discussed in Chapter 2, there is increasing pressure on engineering curricula to shift their focus to topics that are more closely aligned with current research funding opportunities such as nanoscience and biomedical engineering. At the nation's research universities reputations are tied to the success of their research, and both faculty and students are attracted to universities with successful research programs. In engineering colleges, research is the fuel that drives the educational engine. Sponsored research provides support for graduate students and laboratory infrastructure, and it attracts top faculty to the field and the institution. Undergraduate programs benefit substantially from research by the trickle-down effect, whereby undergraduates become engaged in research or take courses that focus on topics that have high research profiles. Faculty develop courses based on their research, and their teaching tends to follow their research. As such, areas of science and engineering that are well supported in research attract the top faculty and tend to drive the educational curriculum on both the graduate and the undergraduate levels. Conversely, areas that are poor in research have fewer faculty, fewer courses, and fewer students. While the merit of a strong correlation between research and education is debatable, its existence is not. Unfortunately, and as discussed in the chapters that follow, corrosion as a discipline is suffering nationally from a paucity of research, of faculty, of courses that incorporate the subject matter, and of students interested in the field. As such, corrosion engineering education in the undergraduate curriculum in many engineering colleges today is minimal. The decrease in DOD funding for research directed to corrosion is an important factor in its declining appeal to faculty and students alike. The result is that the workforce now entering industry and government service has little or no training in corrosion, even though the jobs they are taking require an understanding of how corrosion impacts component design and performance.

This is in contrast to the solid mechanics research community, which has participated thoroughly in developing advanced design approaches to minimize brittle fracture and fatigue failure. The success of mechanics modeling in alleviating mechanical degradation at the design stage has led to it being given a prominent place in the curriculum for many engineering disciplines—for example, aerospace, civil, and mechanical engineering. Success in modeling corrosion could bring important advances in combating corrosion through design and might result in greater emphasis on corrosion engineering in engineering curricula.

The motivation for this study, which focuses on the education of the workforce, current and future, is that knowledge-based and skills-based education is critical to preventing and mitigating corrosion at all stages in the life cycle of a product. The FHWA and NACE International study *Corrosion Costs* made recommendations for preventing or mitigating corrosion.[12] Education is the key to carrying out these recommendations:

- Preventive strategies in nontechnical areas:
  — Increase awareness of the significant corrosion costs and the potential savings.
  — Change the misconception that nothing can be done about corrosion.
  — Change policies, regulations, standards, and management practices to increase cost-savings through sound corrosion management.
  — Improve education and training of staff in the recognition and control of corrosion.
- Preventive strategies in technical areas:
  — Advance design practices for better corrosion management.
  — Advance life prediction and performance assessment methods.
  — Advance corrosion technology through research, development, and implementation.

One key challenge is the decline in student interest in corrosion and in the production of engineers who possess a working knowledge of corrosion. Such a decline threatens to take the field below the "visibility horizon" of both engineering colleges and potential employers of engineering graduates. Corrosion as a subject taught in our tertiary education system is at risk because it is practically nonexistent. The well-documented cost of corrosion to the nation only bolsters the argument that an assessment of corrosion engineering education today is timely and will ensure that the nation not only recovers some of the cost of corrosion but also can rely on the readiness and safety of its critical systems.

Role of Corrosion Research

In line with its charge, the committee has emphasized education, and although it was not specifically charged with examining corrosion research, the committee recog-

---

[12]The study is summarized in Appendix A of this report. For a copy of the report, see http://www.corrosioncost.com/pdf/main.pdf. Accessed April 2008.

nizes that research is one of the key vehicles by which graduate education is effected. Accordingly, it looks at the impact that research has on the development of individual faculty members. Research is critical to understanding corrosion and developing new mitigation strategies. Mainly, however, the committee insists that educating engineers and making them aware of corrosion is the first line of defense in ensuring that corrosion is considered during the design process and over the use lifetime.

Scope of the Study—Metals and Nonmetals

Corrosion, as defined by DOD, the sponsor of this report, is the degradation and loss of function of all materials by their exposure to the environment. Historically, corrosion has meant the destructive oxidation of metals, and this is the way corrosion is often taught. However, the use of materials and the design of new materials are now dominated by nonmetals. Polymers and engineered plastics and composites have been one of the success stories of science over the last 100 years. Between 1980 and 2006, employment in the U.S. plastics industry grew by 1.1 percent per year, and the real value of plastics shipped grew 140 percent, from $114.5 billion to $275 billion.[13] Ceramics, concrete, asphalt, and natural stone remain key components of the public infrastructure, but new hybrid organic/inorganic materials, materials based on nanoscale properties, and biomimetic materials are all increasingly in use. Materials development is becoming an atom-by-atom, molecule-by-molecule, or layer-by-layer construction process. This approach to creating new materials with superior properties is sometimes based on scientific principles, sometimes on combinatorial materials design procedures, and sometimes on imitating nature's own self-organizing processes. Admittedly, some of the combinations of strength, flexibility, and low cost possible with certain metals have not yet been duplicated in new material regimes, nor have the low cost and desirable properties of concrete based on Portland cement been duplicated or significantly improved on. However, the past tells us that new materials free of the weaknesses of current materials will be developed, and many of them will be nonmetallic.

While this report is mainly concerned with corrosion engineering education as it pertains to metals, the committee recognizes that nonmetallic materials such as plastics and composites are increasingly being used in applications that up to a few decades ago were the exclusive domain of metals. Indeed the total production of resins—organic polymers, plastics—is now comparable to that of metals, and of this total, a significant fraction is being used in structural applications.[14] Contrary to some perceptions, however, plastic materials may be susceptible to interactions that degrade their properties (see Box 1-2).

---

[13]See http://www.plasticsindustry.org/industry/facts/usa.pdf. Accessed April 2008.

[14]Alan S. Wineman and Kumbakonam Ramamani Rajagopal, *Mechanical Response of Polymers: An Introduction*, Cambridge, England: Cambridge University Press (2000).

BOX 1-2
Degradation of Polymers and the Education of Engineers

There are many reasons for the growing use of polymers and composites, only one of which is their perceived enhanced resistance to the degradation of their properties by the environment. The inherent degradability of an organic resin must be modified by compounding it with necessary additives. Absent this important step, the performance and safety of the material will be severely compromised. This susceptibility to degradation varies widely according to the chemistry of the polymer and the environment in which it is used. For example, the fundamental carbon–carbon bond is sensitive to solar ultraviolet radiation, especially in the presence of oxygen. Because of this susceptibility, polymers are seldom employed in the neat form, especially for load-bearing applications. The intrinsic properties of a polymer, including its degradability, can almost always be modified by including one or more additives, which can be selected from a wide range to suit the particular circumstance. Such modification can retard the damage caused, for example, by oxidation.

The action of solvents is another reason for property degradation. Solvents may cause undesirable plasticization (lowering the glass transition temperature) and a loss of mechanical strength. Even small amounts of adventitious water, for example, have been known to cause problems in epoxy resins. In addition, certain solvent-polymer combinations may be peculiarly prone to stress crazing and stress cracking. Although polymers in general are relatively immune to biological attack, they are not universally so. Thus aliphatic polyesters are known to be susceptible to certain forms of microbial attack, with negative consequences. To combat these and similar phenomena, a thorough knowledge of polymer properties needs to be acquired by engineers working in the organic materials field.

In structural applications where enhanced mechanical or other properties are required, composite materials have become increasingly important. In the present context, a composite can be understood to refer to a multicomponent system in which a high-modulus fiber (say, carbon or glass) is dispersed in a stress-transmitting plastic matrix, frequently one of the thermosetting class of polymers, such as epoxies. The integrity of the composite depends on the integrity of both the matrix and the fibrous component. Technological advances in such composite materials have led to their increasing use in multiple transportation applications, in industrial and domestic infrastructure, and even in small to medium-size bridges.

While much of the committee's data gathering has been in the field of the corrosion of metals, the committee recognizes that environmentally modulated degradation is a pervasive phenomenon that affects all classes of materials. The growing penetration of organics—principally plastics and composites—into the materials arena means that they are also of concern. Property deterioration in organic materials, commonly referred to as degradation rather than corrosion, is, as with metals, a phenomenon that can be alleviated with currently available science and technology, provided that this knowledge is appropriately applied. The fact that cases of polymer and composite failure caused by some form of environmental interaction continue to occur in significant number suggests that education in this area is entirely inadequate. Many but not all engineering curricula pay relatively scant attention to the properties of organic materials, and the phenomena associated with their degradation are often taught only at a superficial level. Even in undergraduate and graduate programs that are dedicated to organic polymeric materials, education in degradation and its mitigation is decreasing, a situation that parallels what is taking place in the metals field.

The possibility of plastic and composite failure should be a constant concern for design engineers, many of whom have not been taught about this possibility and are therefore far less familiar with this consideration than is desirable. The annual costs of polymer and composite failure due to environmental factors have not been quantified in the same way as the costs of corrosion in metals, but they are certainly significant. As with metallic corrosion, appropriate education of materials engineers and other kinds of engineers in the degradation of plastics and composites must become a tool for dealing with such degradation.

OUTLINE OF THE REPORT

This report assesses the state of corrosion engineering education in the United States and makes a series of recommendations for improving the situation. Chapter 2 summarizes the committee's assessments at the undergraduate and graduate levels and looks at what training and on-the-job education are being offered by industry and government. Chapter 3 looks at the impact of the current status of corrosion engineering education on government and industry. It examines whether meeting government and industry needs demands new approaches in corrosion engineering education. For instance, DOD takes an aggressive stance against corrosion, which directly affects its readiness, but it is unclear whether the nation is producing engineering practitioners who can implement the corrosion strategies of DOD and other national entities. The challenges industry faces with regard to corrosion and the scarcity of professional staff knowledgeable about corrosion make for a difficult situation, and the methods used to cope will be described. Industry challenges include the long-term maintenance and safety of structures, pipelines, and highways. The committee draws conclusions and makes recommendations on the direction the United States should follow as it seeks to reinvent its system for educating the engineering workforce in corrosion engineering education.

2

An Assessment of
Corrosion Education

Corrosion impacts virtually every infrastructure system and manufacturing process and product. The fields of corrosion science and corrosion engineering try to respond to the desire for safe, reliable, economical, and design-lifetime-long performance of industrial and consumer equipment exposed to service environments. The workforce responsible for addressing the corrosion problems faced by both the government and private industry possesses various levels and types of corrosion engineering education. Corrosion technologists are needed to perform repeated crucial functions, such as those of paint inspectors and specifiers, cathodic protection designers, and installers. Well-established practices, such as those developed by the standardization communities, are often put to use in performing these functions. Both practicing and newly minted engineers, who do most of the design work, must possess some degree of corrosion awareness. Sometimes they also need to know enough about materials and corrosion to take corrosion into account in the design process. Corrosion specialists devoted to the selection and implementation of corrosion protection methods or to selecting materials that can withstand corrosive environments are also needed. Finally, there is also a need for a lesser number of experts specialized enough in corrosion fundamentals to investigate new and unexpected corrosion problems, make decisions about them, and act to mitigate the problems. These experts would primarily address novel challenges that cannot be handled with off-the-shelf knowledge or practices, such as the need for a new environment-friendly, corrosion-resistant coating to replace an existing hazardous coating or the need to extend corrosion-limited lifetimes. While they are not needed often, experts are an important part of the corrosion workforce because without

---

BOX 2-1
Knowledge-Based Education and Skills-Based Training

Training, or skills-based education, is focused on imparting a defined set of skills and responses to be applied in a generally known set of conditions. Training generally does not provide a fundamental understanding of the field but teaches how to recognize a condition or situation to select the best solution. Skills-based training is distinct from knowledge-based education in that it does not ultimately give an individual the depth of understanding required to apply a body of knowledge to a situation that has not been previously encountered. During the course of this study, the committee weighed the value of both skills-based training and knowledge-based education.

---

them the challenges would not be overcome. Moreover, those experts who are also educators are the ones responsible for teaching our future corrosion experts.

Two types of education typically go into the formation of this workforce. One is training or skills-based education and the other is knowledge-based education (see Box 2-1). The industry and government panels invited for discussion during the course of this study believe that there is an important role for both training and knowledge-based corrosion education, depending on the job function and desired outcomes. Many corrosion-related functions can be performed by trained corrosion technologists. The corrosion workforce pyramid shown in Figure 2-1 captures the concept that a relatively large number of technologists are needed to support the U.S. infrastructure, including all the sophisticated equipment associated with the country's large industrialized economy. For instance, there are thousands of bridges and thousands of miles of buried pipelines in the United States that require cathodic protection and coatings. In contrast, only one or two engineers specialized in corrosion (identified as "experts" and "specialists" in the pyramid) may be needed for every 100 or more other kinds of engineers in a large company or organization. In the United States, corrosion technologists are often trained by supervising their performance of repeated and predictable corrosion tasks (on-the-job training) or in conjunction with short courses or associate degrees offered by a limited number of community colleges. This education focuses on a defined set of skills and responses to a generally known set of conditions that are often repeated over and over again. A corrosion technologist often implements standardized practices because his or her education generally did not impart the fundamental understanding required to apply a body of knowledge to a situation that has not been encountered before. Such a situation calls for a knowledge-based education.

Knowledge and understanding enable an individual to analyze a new problem and to devise new solutions that go beyond the catalog of known responses to

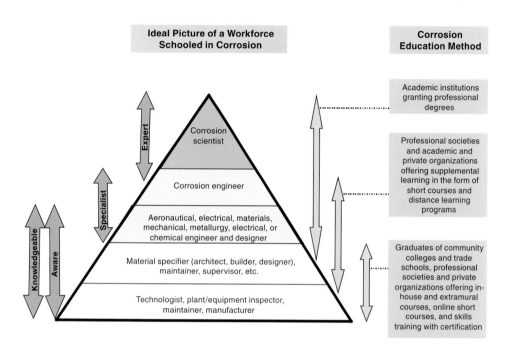

FIGURE 2-1 Corrosion workforce pyramid. The pyramid shows the various categories of corrosion professionals and the knowledge they need—from the large numbers of technologists and other professionals in engineering and related disciplines who would be aware of and knowledgeable about corrosion and its mitigation, the engineers who would be considered corrosion specialists, and a small number of corrosion scientists and corrosion engineers who are experts. The column on the right shows the education paths they typically follow. Broadly speaking, the workforce can be characterized as follows: Technologists, such as paint inspectors and specifiers, and cathodic protection designers and installers perform repeated crucial functions; undergraduate engineering students in materials science and engineering (MSE), who upon graduation should be knowledgeable in materials selection; undergraduate engineering students in other design disciplines, such as mechanical, civil, chemical, industrial, and aeronautical engineering; and MSE graduate students, who should be very knowledgeable in materials selection and in some cases will go on to be experts in the field of corrosion. SOURCE: Adapted from John R. Scully, presented at 16th International Corrosion Conference, Beijing, China, September 2005.

known problems. Therefore, at least some of those responsible for design, manufacture, and product lifetime must be knowledgeable in corrosion and materials fundamentals, so that they are equipped to address corrosion issues arising from the ongoing introduction of new materials and designs. Knowledge-based education is typically delivered through short courses and university-level education. Practicing engineers who focus on the design, manufacture, and processing of components and equipment usually come from one of the traditional engineering fields, such as chemical, civil, and mechanical engineering. They must be aware of the potential problems due to corrosion and be able to recognize when they need to call in a corrosion expert. Such engineers benefit from elective courses in corrosion, short courses, and distance learning in both materials engineering and corrosion science and engineering. Many corrosion specialists learn at the graduate level and carry out their thesis or dissertation research in a university setting; others learn through a lifetime of on-the-job experience and short courses.

In summary, the corrosion workforce is educated by means that span a broad educational spectrum:

- Bachelor's and associate's degrees awarded to corrosion-aware and corrosion-knowledgeable engineers and corrosion specialists.
- Graduate education to produce corrosion experts.
- On-the-job training, continuing education through distance learning, and short courses to produce corrosion technologists, corrosion-aware engineers, as well as corrosion specialists and experts, depending on the course and the starting knowledge of the student.

UNDERGRADUATE CORROSION EDUCATION

At workshops convened to gather information, the committee heard from panelists representing various industrial and governmental sectors that their respective employee pools displayed very different levels of corrosion awareness. This is hardly surprising; given the vast body of engineering knowledge necessary to operate modern industries and agencies, not all engineers should be expected to have a mastery of corrosion. In general, however, employers expect engineers holding a baccalaureate with a major in materials science and engineering or metallurgical engineering to have a deep enough understanding of corrosion at a sufficiently fundamental level that they can avoid obvious pitfalls in materials selection and know when to consult corrosion specialists or experts. In contrast, engineers holding a baccalaureate in a nonmaterials field would not be expected to have much understanding of corrosion but could at least be expected to be aware of corrosion. The committee was told that the skill sets of many (although certainly not all) practicing engineers fell short of these basic expectations. This deficiency might reflect inadequate exposure to corrosion in the undergraduate curriculum, ineffective instruction, or even the failure on the part of engineers to remember what they had been taught.

It was apparent at the workshops that the majority of participants thought of corrosion principally in terms of metallic corrosion, occurring by electrochemical mechanisms. Many corrosion classes also focus on metallic corrosion. This chapter, although it, too, focuses on metallic corrosion, touches on nonmetallic corrosion (see, for instance, the discussion in Box 2-2). Two questions arise when assessing

BOX 2-2
Education in the Degradation of Nonmetals

Most contemporary MSE undergraduates will have had some exposure to the properties of organic materials. However, few MSE curricula in this country provide comprehensive instruction in polymeric and composite materials. Historically, the most comprehensive education in polymers and composites is offered by a relatively small number of specialized departments, many outside of engineering colleges. Some of them offer both undergraduate and graduate programs, while others focus solely on graduate education. Even in such comprehensive programs, polymer degradation and failure are rarely primary academic topics nowadays. In the past, the mitigation of polymer degradation and failure by compounding resins with appropriate additives was of great interest in both teaching and research. Paralleling the situation in metallic corrosion, however, funding for research on degradation and related topics has shrunk significantly, leading to decreased faculty interest and diminished treatment of this topic in polymer- and composite-focused curricula. The overall result is that few current engineering graduates will have had any significant exposure to the subject of polymer or composite degradation. While graduate engineers will therefore be very capable of monitoring the change in properties of a polymer in service, they will have no familiarity with or understanding of the interactions in a particular environment/polymer system, and they are unlikely to be able to select a polymer/additive compound or a composite. Ideally, engineers engaged in such a selection, which often involves the substitution of a polymer or composite for a metal, should be familiar with the advantages and shortcomings of both classes of material. It is regrettable in terms of societal costs and public safety that the present educational system rarely, if ever, imparts such comprehensive expertise.

corrosion education from the supply side. First, which types of courses expose students to corrosion, and how comprehensive is that treatment? Second, who takes those courses? At the undergraduate level, corrosion is typically taught in three broad categories of courses.

The Dedicated Corrosion Course

The first and most comprehensive of these is the dedicated corrosion course, typically involving about 40-45 hours of classroom instruction that may also be taken by starting graduate students. A typical modern class covers the fundamental thermodynamics and kinetics of corrosion, the eight forms of metallic corrosion (uniform, galvanic, crevice, pitting, intergranular, selective leaching, erosion corrosion, and stress corrosion), the environmental degradation of nonmetals, and corrosion protection strategies such as coatings, inhibitors, and cathodic protection. The coverage is primarily theoretical, grounded in the theory of corrosion and the principles of electrochemistry. A hypothetical syllabus for such a course is shown in Box 2-3. Another style of dedicated corrosion class is more deeply rooted

---

BOX 2-3
Hypothetical Syllabus for an Undergraduate Course on Corrosion

1. Introduction
    - Importance of corrosion
    - Forms of corrosion
    - Basic electrochemistry: pH, balancing electrochemical equations
        - Anodic, cathodic half-cell reactions
        - Faraday's law
        - Ions in solution, ion transport, current flow
        - Structure of electrochemical interface, potential
        - The four requirements for a corrosion cell
2. Thermodynamics
    - Review of free energy, activity
    - Electrochemical free energy
    - Standard potential
    - Electromotive force series
    - Simple electrochemical cells, cell potential, reaction direction, spontaneous reactions versus forced
    - Nernst equation
    - Effect of concentration on electrochemical cells
    - Reference electrodes
    - Pourbaix diagrams
        - Oxygen reduction and evolution
        - Water, proton, hydronium reduction, evolution, stable region of water
        - Metal: passivity, immunity, corrosion
        - Effect of other oxidizers: chlorine, peroxide, nitric acid
        - Sample diagrams: Fe, Al, Cu, Cr diagrams
        - Use of Pourbaix diagrams
        - Estimated effect of alloying
3. Kinetics
    - Driven and driving systems, electrode polarity
    - Exchange current density
    - Activation polarization, Butler Volmer equation
    - Tafel equation
    - Mixed potential theory: redox reaction, coupled reactions—a corrosion cell, corrosion potential, and current density
    - Simple Evans diagram
    - Effect of added oxidizing agent
    - Concentration polarization

---

in learning practical skills, treats the eight forms of corrosion more descriptively, works with case studies, and teaches some design issues and corrosion remediation strategies. Other courses dedicated to corrosion might link it to batteries and fuel cells, where corrosion is providing electrical power. A comprehensive electrochemical engineering approach might cover the same fundamental principles

Transport limitation of cathodic reactant, effect of flow or stirring
Effect of oxygen transport limitation
Transport limitation of anodic reaction
4. Measurement of Corrosion Rate
    Mass loss, mass loss rate, penetration rate
    Measured polarization curve and underlying Evans diagram
    Potentiodynamic polarization, Tafel extrapolation, fit to equation
    Linear polarization, Rp
    Experimental considerations: sample, cell, electrolyte, RE, CE, etc.
    Atmospheric corrosion tests
5. Corrosion Phenomenology
    Uniform (examples of Fe, Al compared with pH, NaCl, etc.)
    Galvanic corrosion (using Evans diagrams to explain)
    Erosion corrosion and fretting
    Passivity, stainless steel alloying (other brief examples)
    Pitting
    Crevice corrosion
    Intergranular corrosion
    Dealloying
    Environmentally assisted cracking
        Stress corrosion cracking
        Hydrogen effects
        Corrosion fatigue
6. Corrosion Prevention
    Materials selection, alloy corrosion characteristics
    Coatings
    Inhibitors
    Cathodic protection, sacrificial and impressed current
    Anodic protection
7. Special Materials/Environments
    Polymer corrosion/degradation
    Atmospheric corrosion
    Oxidation
    Underground corrosion
    Rebar in concrete
    Microbial effects

of corrosion along with other electrochemical applications, such as chloroalkali synthesis, electrodeposition, and electrowinning. After taking one of these courses, an engineer should have a strong enough foundation that, after on-the-job training, he or she will be able to avoid design blunders and recognize when his or her knowledge limitations necessitate calling in a specialist or expert.

Survey Course That Includes Corrosion

Few undergraduate materials science and engineering (MSE) programs in the United States and even fewer nonmaterials engineering programs offer (much less require) such a dedicated course. Many programs provide the second category of course—that is, an overview of corrosion in classes required for all students. One approach in this category is an introductory, survey-type course offered early in an undergraduate curriculum. Typically this would be an introductory materials science course taken by all materials majors or by students in mechanical, civil, and other engineering fields.[1] Other schools might cover corrosion in a course on the mechanical behavior of materials. Most textbooks for this type of course present corrosion at the back of the book, adding some elementary electrochemistry to build on a foundation of thermodynamics and physical metallurgy gained earlier. Typically, a single lecture is devoted to corrosion, although, unfortunately, some instructors might not make it all the way through the textbook. Assuming that the student did attend the lecture, he or she is likely to know that corrosion requires an anode, a cathode, electrical contact, and ionic contact. For dissimilar metal couples, the engineer may be able to consult a handbook on the galvanic series and identify which metal would act as the anode in service and which as the cathode. To give an idea of the expectations for engineers, the only such material covered in the engineering license fundamentals of engineering (FE) exam for professional engineers offered by the National Council of Examiners for Engineers (NCEES) is the electromotive force series. A graduating engineer might also have access to a corrosion report prepared by an expert that would allow him or her to make decisions or reach conclusions. She or he is unlikely to be able to proactively address specific corrosion problems in design or independently analyze corrosion failures encountered on the job.

Senior Design Course

The third kind of course where a student might gain some corrosion awareness is the senior capstone or design course; here students are expected to synthesize the knowledge acquired in many different courses to tackle a particular design problem.[2] For instance, such a course for a materials engineer would focus on the selection of materials appropriate for specific applications, so the student engineer would have to consider the impact of corrosion, along with other factors, on the

---

[1]Although the focus of this report is engineering education, the committee notes that often some electrochemistry and corrosion are taught in freshman chemistry classes.

[2]A capstone course is a course offered in the final semester of a student's major. It ties together the key topics that faculty expect the student to have learned during the major, interdisciplinary program, or interdepartmental major.

functional success of the project. If that student has had little or no exposure to corrosion, a course whose objective is to help the student synthesize knowledge already mastered is unlikely to teach anything else than that corrosion could take place and degrade system performance.

Discussion

Unfortunately, while these three categories of corrosion education are available in some schools, many students, particularly those in fields of engineering other than materials, are likely to graduate with no formal exposure to corrosion science or engineering. This situation explains the limited corrosion-related skill sets that students are bringing to the workforce.

Given the enormous financial and strategic importance of corrosion, as discussed in Chapter 1, how is it that most U.S. engineers can graduate with so little grasp of corrosion? In the committee's opinion, the answer lies in the growing number of competing topics that the graduates must master. As engineering becomes increasingly complex and interdisciplinary, there is constant pressure to keep adding fresh material to the curriculum, including courses on new tools that lead to a deeper understanding of all materials while keeping the course load to a total of 120-128 credits. (Examples are computational tools for modeling and visualizing everything from bonding to structure formation to macroscopic processes.) This pressure comes from various stakeholders: students, who want to be competitive for employment or admission to graduate schools; faculty members, who sincerely believe that every well-educated student should know a reasonable amount about his or her own research specialty; and also employers, who want new graduates to be conversant in the latest findings. Most engineering educators recognize that curricula are already saturated and accept that if new topics are to be added, old ones must be subtracted or diluted.

Corrosion education tends not to fare well in the face of these pressures. Despite its importance, corrosion is not new, and few consider corrosion science and engineering to be at the cutting edge. The very thought of corrosion can be off-putting to students,[3] who feel that they should be learning about new technologies with the potential to change the world. Few engineering educators and students grasp the wealth of strategies that are available to prevent corrosion and would rather spend class time on topics that they perceive as more useful.

Results from a questionnaire circulated to engineering educators in conjunction with the present study tend to support this view (see Box 2-4 for a discussion

---

[3]A couple of panelists thought that corrosion would be taken more seriously if the name were changed. The corollary cited was the term "tribology," which has come to be used in place of "wear."

BOX 2-4
Data on U.S. Corrosion Education[1]

Engineering schools in the United States were asked to complete an online questionnaire so that the committee could obtain a clearer picture of the status of undergraduate corrosion education. Questionnaires were sent to 83 educational institutions that included, but were not limited to, all ABET-accredited MSE programs and the top 20 engineering schools in terms of numbers of bachelor's-level graduates produced annually. Thirty-one responses were received (37 percent response rate); 19 of them were from MSE programs and 12 were from other programs (chemical, civil, mechanical, environmental, or general engineering). Three of the respondents followed a quarter system; the remaining 28 a semester system. The committee recognizes that there is likely to be a strong bias in this exercise, since schools participating in corrosion education would be more likely to respond than schools that did not. Accordingly, the results should be viewed as giving the best case for corrosion offerings rather than a truly accurate picture.

Nineteen of the respondents offered a course or courses specifically on corrosion, identifying a total of 26 courses altogether, 16 of which are offered every year. Of these, only 6 undergraduate courses were identified as "required." The remaining 20 courses comprised 6 elective undergraduate courses, 7 elective graduate-level courses, and 7 elective courses aimed at a mixed audience of undergraduates and graduate students. Although 6 courses were identified as "required," only one program indicated that it required a corrosion class, along with two specializations within programs. Reasons cited for offering corrosion courses included student interest; the belief that it was essential for materials/metallurgical scientists and engineers to know about corrosion and important for many different careers; and employer demand.

Eleven respondents offering corrosion courses indicated that students from other departments took the corrosion course; these students had a wide range of engineering backgrounds. All of the corrosion courses taught the electrochemical fundamentals of corrosion and ways to minimize corrosion by design. Some focused on metallic corrosion, while others covered the degradation of a broad spectrum of engineering materials, including metals, ceramics, polymers, wood, biomaterials, and biodegradable materials in many different service environments. At some schools the content of corrosion courses had changed significantly in the last 10 years: Many reported broadening the range of materials covered, others had increased their emphasis on electrochemical mechanisms in metallic corrosion, and a few emphasized the role of corrosion in fracture mechanics. The questionnaire process identified only three laboratory-based corrosion courses and three corrosion courses offered as distance courses. It is likely that

of the questionnaire). More than half of the respondents (19 of the 37) said that their institution offered a specialized course in corrosion.[4] However, it is essential

---

[4]Survey recipients were asked about corrosion education at their institution, but the survey did not specify whether corrosion referred exclusively to metallic materials, to the low-temperature degradation of metallic materials, or to general materials degradation. The majority of responses confirmed the committee's expectation that most individuals consider corrosion to be the degradation of metallic materials.

these trends reflected the individual strengths of instructors; the survey revealed that corrosion courses were taught by instructors with a wide range of expertise and experience, from corrosion specialists with active research programs in corrosion science and engineering through those with related expertise in electrochemistry and applied chemistry, to those with no formal training in corrosion. Twelve respondents did not offer a specialist course in corrosion. Of the schools that did not offer a specific course, three indicated that other topics had higher priority, four indicated that they had no one to teach such a course, and five said that corrosion was covered in other courses. Seventy-nine percent of all respondents indicated that corrosion was covered in other courses. These other courses were most commonly an introductory materials course required of students in materials science and engineering. Corrosion was also treated in some thermodynamics, design, chemistry, processing, and mechanical behavior courses. Five schools required mechanical engineering undergraduates to take a course that included corrosion, and one or two schools required students in industrial engineering, chemical engineering, civil engineering, manufacturing, and general engineering to take such a course. Students taking classes in which corrosion was covered along with other topics represented a range of engineering majors, along with students studying physics and dentistry.

Because the lack of qualified instructors had been widely cited at workshops as a reason for not offering corrosion courses, the questionnaire asked whether the responding school would consider hiring a faculty member whose technical focus was corrosion. Fifty-eight percent of respondents replied that they would consider making such an appointment, while 42 percent would not. Of those that would consider such an appointment, about half would appoint someone to replace a retiring faculty member. Most of the remainder would consider such an individual provided that they were competitive with candidates across a broad range of other technical areas and had broader expertise. Of those schools that would not consider appointing a corrosion expert, 91 percent believed that other topics had a higher priority and 9 percent believed the availability of research funding was limited. Respondents were asked to identify where graduates from their undergraduate programs eventually found employment. Averaged across all responses, 24 percent ended up working in design, 41 percent in manufacturing, 23 percent in research or academia, and 12 percent in other areas.

---

[1] Summary of the results of the questionnaire, which are reported fully in Appendix B.

to note that personnel at institutions participating in corrosion education might be more likely to respond to the questionnaire, so that the results should be viewed as probably overstating the real situation.[5] Some schools also required the corrosion course for those majoring in disciplines such as materials science, materials science/mechanical engineering joint degree, metallurgy, and chemical engineering,

---

[5]According to data presented to the committee, a DOD survey of schools found that of the 72 institutions surveyed, 31 offered a corrosion course.

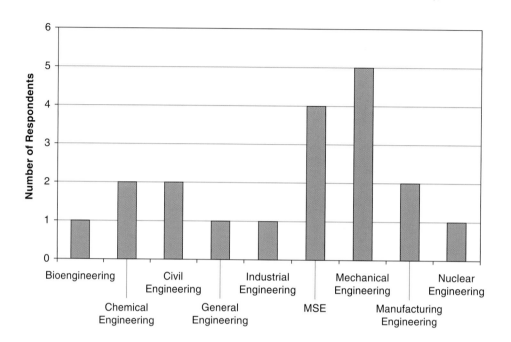

FIGURE 2-2 The number of respondents to the committee's questionnaire who indicated a corrosion course was required for a particular major.

as well as the biomaterials and metals specializations within materials science and engineering.

A few of the schools that responded to the committee's questionnaire offer interested students a dedicated corrosion course as an elective. Of the schools that did not offer a dedicated course in corrosion, a quarter said they placed a higher priority on other topics while a third stated that their institution did not have anyone with appropriate training to teach a specialized corrosion course. Others seemed to feel that the coverage of corrosion provided in other courses was adequate. In other words, it seems to the committee that corrosion education is not commonly considered to be a crucial component of an engineer's professional education. Figure 2-2 shows data from the committee's questionnaire on the number of responding institutions' required corrosion courses. Figure 2-3 shows the most frequently cited reasons for not offering corrosion courses. Figure 2-4 shows the availability of corrosion courses in the top 10 Ph.D-granting institutions and the top 10 4-year engineering colleges.

Those schools offering a dedicated corrosion course reported that it was being taken by students from a wide range of engineering majors and by others as well,

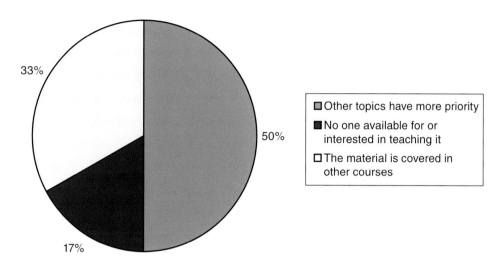

FIGURE 2-3 Most frequent reasons for not offering a corrosion course.

such as dentistry students. This suggests that there is, indeed, a demand from students for good, high-quality dedicated courses when they are offered by engaging faculty. At one school that emphasizes corrosion, the undergraduate corrosion course has been expanded to include corrosion batteries and fuel cells to attract students. Nevertheless, only about 8 percent, 5 percent, and 12 percent of mechanical, civil, and chemical engineers, respectively, took this course. Data on enrollment in corrosion courses at two schools are shown in Figure 2-5. One would expect these statistics to reflect a best-case scenario, given the strength of corrosion instruction at these schools. One can only conclude that few undergraduate engineers take corrosion classes, even when good ones are available. Furthermore, the time spent on the topic in courses that "cover" corrosion amounts to only a very small fraction of the overall discussion time (see Figure 2-6).

The committee is aware of two exceptions to the trend of little formal corrosion education at the undergraduate level. Kilgore College in Texas offers an associate of applied science degree in corrosion technology, with an emphasis on applications in the oil industry.[6] In addition, the University of Akron in partnership with NACE is planning a B.S. in corrosion engineering. There are also plans to submit the Akron program to ABET, Inc. (formerly the Accreditation Board for Engineer-

---

[6]Kilgore's corrosion technology associate's degree was established in 1980. In 2007 the program had 70 students. SOURCE: Kathy Riggs Larsen, "Wanted: Corrosion Professionals," *Materials Performance*, December 2007.

FIGURE 2-4 Availability of corrosion courses in the top 10 Ph.D.-granting institutions and the top 10 4-year engineering colleges. *Right*: materials science and engineering curricula. *Left*: mechanical engineering curricula. Data on the availability of corrosion classes was based on a 2004 survey of online descriptions of curricula. Of the 20 schools examined, only three materials departments required a materials selection course and one required a corrosion course. Five departments taught corrosion as part of another required course, three taught it from the standpoint of materials section, and two had a focus on corrosion mechanisms. Only 1 of the 20 mechanical engineering departments requires a materials selection course and 3 required a corrosion course. Six schools taught corrosion as part of another required course, three teaching it from the perspective of materials selection and three focusing on corrosion mechanisms. *Based on *U.S. News and World Report*'s 2004 listing of top 10 schools in each category. †Materials selection covered as part of design and/or materials courses. SOURCE: David H. Rose, DOD Reliability Information Analysis Center.

Left: mechanical engineering curricula

Top 10 Ph.D.-Granting Engineering Universities*	Materials Selection Course Required?	Corrosion Course Required?	Corrosion Taught as Part of Another Required Course?	Mechanistic (M) or Materials Selection (MS) Based Curricula?
1 MIT	N†	N	N	—
2 Stanford	N	N	Y	M
3 University of California (Berkeley)	N†	N	N	—
4 Caltech	N	N	N	—
5 George Institute of Technology	N†	N	N	MS
6 University of Illinois	N†	N	Y	—
7 University of Michigan	N†	N	N	M
8 Carnegie Mellon University	N	N	N	—
9 Cornell University	N†	N	N	—
10 Purdue University	N†	N	N	—
Top 10 4-Year Engineering Colleges*				
1 Embry-Riddle	NA	NA	NA	NA
2 U.S. Air Force Academy	N	Y	N	—
3 St. Louis University-Parks College	N	Y	N	—
4 U.S. Naval Academy	Y	Y	Y	M,MS
5 Rose Hulman Institute of Technology	N	N	Y	MS
6 Cooper Union	N	N	Y	—
7 Bucknell	N	N	Y	—
8 U.S. Military Academy	N	N	N	—
9 Cal Poly-San Luis Obispo	N	N	Y	M
10 Harvey Mudd College	NA	NA	NA	NA

Right: materials science and engineering curricula

Top 10 Ph.D.-Granting Engineering Universities*	Materials Selection Course Required?	Corrosion Course Required?	Corrosion Taught as Part of Another Required Course?	Mechanistic (M) or Materials Selection (MS) Based Curricula?
1 MIT	Y†	N	Y	MS
2 Stanford	N	N	N	—
3 University of California (Berkeley)	N	Y	Y	M,MS
4 Caltech	N	N	N	—
5 George Institute of Technology	Y†	N	Y	MS
6 University of Illinois	Y	N	N	—
7 University of Michigan	N	N	Y	—
8 Carnegie Mellon University	N	N	N	unk
9 Cornell University	N	N	N	—
10 Purdue University	N	N	N	—
Top 10 4-Year Engineering Colleges*				
1 Embry-Riddle	NA	NA	NA	NA
2 U.S. Air Force Academy	N	N	N	—
3 St. Louis University-Parks College	N	N	N	—
4 U.S. Naval Academy	N	N	N	—
5 Rose Hulman Institute of Technology	N	N	N	—
6 Cooper Union	N	N	N	—
7 Bucknell	N	N	N	—
8 U.S. Military Academy	N	N	Y	—
9 Cal Poly-San Luis Obispo	N	N	N	M
10 Harvey Mudd College	NA	NA	NA	NA

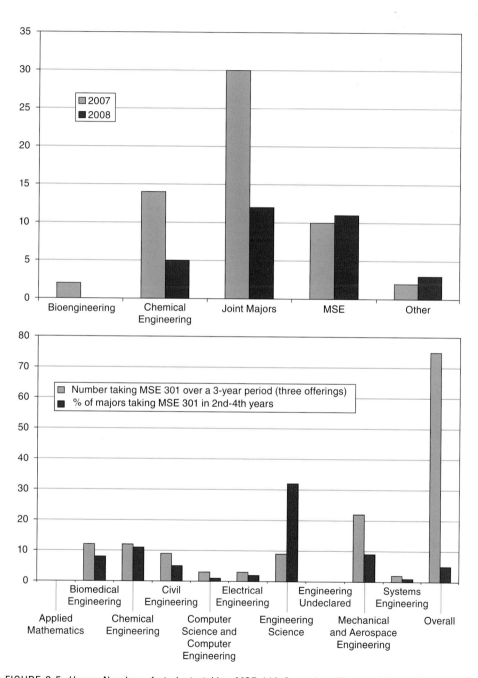

FIGURE 2-5 *Upper:* Number of students taking MSE 112 Corrosion (Chemical Properties) at the University of California, Berkeley, in 2007 and 2008. *Lower:* Number of students from each major taking MSE 301 at the University of Virginia over a 3-year period and the percentage from each major taking that class.

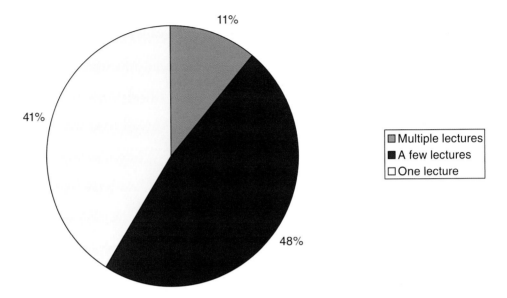

FIGURE 2-6 Number of lectures on corrosion when corrosion is covered. Data are for all schools responding to the committee's survey.

ing and Technology) for accreditation. (See Box 2-5 for a discussion of ABET.) This program is aimed at bridging the gap in the workforce between individuals with an associate's degree in corrosion and those with graduate degrees. While it is too soon for the committee to draw any conclusions about the Akron program, it will be worth watching over the next several years.

To assess the demand for expertise in corrosion, at its second and third meetings, the committee heard from industrial and government agency panels whose members represented a broad spectrum of organizations that employ engineers. Few employers mentioned any need for corrosion technicians at the associate's level or corrosion engineers at the B.S. level. Instead, most of them valued employees at these levels who brought a broader skills set to the workplace, so that they could tackle a variety of projects and tasks. Employers expressed more concern about the lack of fundamental knowledge (e.g., thermodynamics) than about the lack of corrosion knowledge among their B.S.-level engineers. The overarching concern was that on many occasions those making design decisions did not realize that they did not know anything about corrosion. The employers appear to want all engineers making design and materials selection decisions to have enough exposure to corrosion to realize that they do not know enough about it to make the decision alone and that they need to consult a specialist.

BOX 2-5
ABET and Accreditation

ABET, Inc. (www.abet.org) provides accreditation for programs in engineering and technology. ABET also accredits programs in computer science and in the applied sciences. ABET has a worldwide presence and is recognized by the Council for Higher Education Accreditation. Because its membership comprises 28 professional societies, it is the professions, as stakeholders in the quality of ABET's accredited programs, that help to assure educational quality.

In 2006, the latest year for which data are available, ABET accredited 1,787 engineering programs at 364 institutions and 670 technology programs at 226 institutions. A program is an academic course of study leading to a degree and is not the same as a department, which is an administrative unit. Because many universities have both engineering and technology programs, the number of universities visited by ABET is not additive. Within those programmatic categories for which there are program criteria, in 2006 there were 75 engineering accredited programs in materials-related subjects (ceramics 7, materials 58, and metallurgical 10). Also, the subject category (say, metallurgy) does not necessarily match the name of the degree (say, materials science).

For a program to obtain and maintain accreditation it must be reviewed against published criteria. A program's self-evaluation is reviewed by an ad hoc team of specialized and trained peers, who then visit the program to verify the program's own report. If all is well, accreditation is granted for 6 years. Programs with weaknesses may be visited more frequently, and in the rare case that it has deficiencies when it is measured against the ABET criteria, the program may be asked to show cause—that is, say why accreditation should not be removed. No show-cause action is undertaken without giving the program an opportunity to cure its deficiencies and carry out a new self-evaluation.

ABET's general criteria define the minimum educational requirements for a graduate to be deemed an applied scientist, computer scientist, engineer, or technologist. Its program criteria are written with the assistance of the relevant professional society or societies, and these criteria are used to judge programs with specific modifiers. For example, the American Society of Civil Engineers suggests draft program criteria to ABET for all programs in civil engineering. New program areas or areas with a limited number of programs may not have their own criteria, so they are judged against more general criteria by evaluators familiar with the objectives of that program. Some program areas, such as materials, have more than one society dedicated to their discipline; in this case, one society is designated as a lead society for the particular discipline and the others are called "cooperating societies." The materials societies represented in ABET include the National Institute of Ceramic Engineers (NICE), The Minerals, Metals and Materials Society (TMS), and the Materials Research Society (MRS), which is an associate member.

Beginning in 1997, ABET modified its criteria to emphasize continuous improvement, definition of objectives, and assessment of outcomes (the abilities of graduates). Typically, today's criteria contain eight categories: students, program education objectives, program outcomes, continuous improvement, curriculum, faculty, facilities, and support. If program criteria exist, then they are listed as the ninth category of criteria. A significant contribution from professional societies to the program criteria is the coherent list of educational outcomes expected from a program's graduates at the time of graduation. Because of ABET's approach, the committee spent some time developing educational outcomes for each of the many levels of education that might produce a worker in corrosion control (see Appendix F).

These desires are modest but are apparently not being satisfied. How, then, could engineering programs do a better job of ensuring that their graduates know when they are reaching the limits of their knowledge on corrosion and understand that they should consult experts? This would seem to demand more awareness and appreciation of corrosion than is currently imparted at most engineering schools.

More attention to graduate programs that emphasize rigorous, relevant corrosion science and engineering could increase the supply of faculty capable of instituting corrosion programs. Realistically, however, there would not appear to be sufficient funding for every undergraduate program to have its own corrosion specialist. Fortunately, the committee sees opportunities to leverage the expertise of faculty members engaged in complementary research on, say, the applications of electrochemistry (ranging from battery research to work on chemical mechanical planarization) or the mechanical behavior of materials. Alternatively, members of the chemistry faculty or even the MSE faculty (not corrosion specialists) could also teach undergraduate corrosion courses, provided appropriate teaching materials were available. Since most engineering curricula require students to take technical electives, there is an opportunity for nearly every engineering program to offer a course on designing against corrosion.

GRADUATE CORROSION EDUCATION

Graduate education in MSE is the most direct way to produce corrosion specialists, those at the top of the corrosion workforce pyramid. These corrosion specialists, in the committee's view, are the engineers who can use the fundamentals of corrosion science and engineering to address difficult, out-of-the-ordinary corrosion problems and to advance the field by creating new knowledge, techniques, and instrumentation. Graduate MSE education occurs mostly in MSE departments but also, at times, in a corrosion group or center within a chemical or mechanical engineering department.[7] By extension, graduate corrosion education takes place in these departments as well as, in the committee's experience, in civil engineering departments.

Typically, a graduate student becomes knowledgeable in a particular engineering field by taking a sequence of classes and doing research on a focused topic, leading to a master's or Ph.D. thesis. Graduate engineering education usually involves approximately eight classes at the M.S. level or 12-15 three-credit classes at the

---

[7]There are 100 MSE, 223 mechanical engineering, and 142 chemical engineering programs in the United States. Of the 365 programs in mechanical and chemical engineering, 72 percent of the mechanical engineering and 85 percent of the chemical engineering departments house graduate programs.

Ph.D. level.[8] A typical M.S. in the United States is earned in 2 years while a Ph.D. is earned in 4 or 5 years. Students at some schools can also earn a course-based master's degree in engineering by taking a few more graduate courses without conducting any graduate research.

Employer demand for these students after they graduate comes from academia, government, and industry, not least because a new material can be crucial to the mission of any of them, particularly high-tech industries like those manufacturing aircraft engines or nuclear power plants. Of most interest in the context of this study is that many of these industries look for master's- or Ph.D.-level graduates who have some corrosion expertise. One challenge for the graduate corrosion education system is to produce engineers with sufficient fundamental and technical knowledge and good enough critical assessment and communication skills to allow them to contribute immediately to the industry or government organization that recruits them, even though they may lack experience specific to that employer. In light of these considerations, how many master's and Ph.D. students is enough?

According to opinions conveyed to the committee during the panels convened for its meetings, large industrial and governmental organizations need roughly 1 in 50 of their engineers to be knowledgeable about corrosion in order to put together an effective design team. Absent such expertise, an expert consultant must be brought in from outside.

These employers report that individuals with such preparation are usually capable of making substantive contributions to the company's work immediately but also must spend their first 2-3 years in the organization integrating their skills and knowledge with the business culture and the operations and technical application areas of their employer. An alternative to recruiting people with advanced degrees in corrosion is to cultivate them internally. Other approaches are to hire experienced corrosion experts from another company or to use consultants or contract research outfits to solve problems and deal with new challenges as they arise. None of these approaches is as effective as recruiting a freshly minted or experienced corrosion specialist, since it is estimated that it could take over 5 years to develop a corrosion specialist internally.

One industry panelist from the energy sector declared that every materials specialist in his sector should have some level of corrosion education. That panelist also said that corrosion is a key issue in almost every engineering decision involving materials. Panelists from other sectors argued that although only a few corrosion experts are needed in a large company with hundreds of engineers, the experts are crucial nonetheless. A representative of another company argued that corrosion specialists must be capable of performing duties other than those related to

---

[8]Three credits entail 15-16 weeks of course delivery with 3 lecture hours a week in a school that operates on the semester system.

corrosion, because otherwise they would not be fully utilized. A few government sector panelists also said that one corrosion specialist is enough for every 25-100 design engineers; another panelist said that his organization has no in-house corrosion expertise but relies on outside corrosion experts. Another option would be to offer an in-house short course on corrosion to enhance the corrosion awareness of staff.

The committee estimates that currently 19 or so corrosion specialists are graduating each year in the United States from graduate institutions with faculty focusing on corrosion.[9]

The responses to the questionnaire referred to in the preceding section on undergraduate education indicated that of the 31 responding universities, 7 offer mixed (graduate/undergraduate) classes in corrosion and 7 offer graduate classes in corrosion in an MSE department or in a closely related field. This means that approximately 1 in 5 of the responding graduate institutions offered classroom education in corrosion. The frequency with which these courses were offered varies from every year to every other year. The courses are usually but not always taught in MSE departments. They usually cover a mix of fundamentals and ways to minimize corrosion. At the graduate level, corrosion is sometimes covered in courses on thermodynamics, transport, and surface science; 9 out of 44 classes listed had corrosion as a subtopic. Graduate engineering students in MSE are typically not required to take a course in corrosion, and such courses are often offered by departments where there is a faculty member with expertise in corrosion, again mirroring the undergraduate situation.

Corrosion is either a course on its own or a part of a structure-property course that can be taken as a graduate-level technical elective in much the same way fracture and other technical electives are offered. In some cases the graduate-level class in corrosion has a mixture of undergraduate and graduate students and the class is taken by both advanced undergraduate students and new graduate students who might be matriculating from a school or program that did not offer an undergraduate corrosion class. However, in other cases, separate sets of corrosion courses are offered to graduates and undergraduates.

Graduate students in MSE not specializing in corrosion do not necessarily take a graduate corrosion class. This situation mirrors undergraduate MSE education,

---

[9] This estimate is based on the existence of 30 smaller schools, each with one corrosion faculty, that have ongoing research in corrosion and produce 0.5 graduates each year ($30 \times 0.5 = 15$) and 5 large schools with a concentration in corrosion and two faculty members that each produce one graduate ($5 \times 2 \times 1 = 10$), for a total of 25 graduates. About 75 percent, or 19, of these graduates find employment as corrosion experts because 25 percent are employed in other sectors such as microelectronics. This creates a supply of approximately 19 corrosion experts per year by the graduate education route. Another estimate, 24-37 corrosion experts per year, was based on corrosion faculty in technical societies. There may be additional experts from international educational programs.

where the availability of undergraduate classes on corrosion depends on the incli-
nation and technical interests of the faculty in that particular school and the number of technical electives available. In most cases, corrosion is not in the graduate
core curriculum. No matter the engineering field, if no faculty member conducts
research in corrosion or a related field, a graduate corrosion course is unlikely to
be offered. Since the conduct of research in corrosion is heavily contingent on the
availability of research funding—typically obtained through some combination of
grants from the federal government and industry—this factor can in effect dictate
whether graduate students have a chance to study corrosion.

Students who are exposed to a curriculum with a graduate or mixed undergraduate/
graduate corrosion class, or students undertaking an advanced or Ph.D.-level corrosion class, as well as focused dissertation research on a corrosion-related topic,
learn a lot about materials fundamentals (structure, thermodynamics, and kinetics
of solid materials), metallurgy, and one or two other related areas, such as materials
characterization, fracture, or surface science. A single class in polymers, composites,
ceramics, and electronic materials (or combinations thereof) teaching structure/
composition/properties relationships may also be taken. Related technical electives
such as computational modeling of the structure and/or deformation of materials,
materials processing, mathematics of materials science, as well as probability and
statistics often round out such a graduate curriculum.

Students are also prepared in the fundamentals of corrosion by undertaking
research, which offers an opportunity to learn about needs, gaps, and research
opportunities by undertaking a literature survey. The research itself may lead to
a thesis, a paper published in a technical journal, presentation of the results at a
national or international symposium, and even to a career decision to carry on with
the research. Students learn to plan, conduct, and analyze corrosion experiments,
perhaps incorporating 10-15 percent modeling content; this leads to discussion
of results and often the making of recommendations. Research may involve the
mechanisms of corrosion and its mitigation and prevention, but there is usually
limited opportunity for materials selection and design. The research can be of either
an engineering or a scientific nature.

Graduate education in a department other than MSE can also produce a
corrosion specialist out of a student who learns a closely related topic such as
electrosynthesis or fuel cell catalysis, to name just two. However, while such
graduates may lack a substantial background in MSE because the courses making
up their degree did not cover the engineering fundamentals of materials in
depth, they often do take a course in electrochemical engineering or theoretical
electrochemistry, where the fundamentals of metals corrosion are often covered.
Faculty in these departments might include people who study time-dependent
degradation of materials properties, and faculty in chemical engineering might
include those studying batteries, fuel cells, or the electrochemical synthesis of

materials and chemicals. This situation somewhat resembles the undergraduate scene described earlier.

Information on the number of university faculty who consider themselves corrosion specialists can be found on the Web site of NACE International (originally the National Association of Corrosion Engineers). The list there includes about 81 faculty members (excluding retired faculty) who teach graduate-level corrosion in their department. One estimate is that 48 of them are active in research and that each produces 0.5 to 0.75 corrosion experts with advanced degrees every year, or 24-37 individuals. Of this number, it is estimated that only about 75 percent are employed in corrosion-related jobs, with the remainder finding other engineering or technical jobs.

Many universities can also identify between 5 and 9 other faculty spread throughout departments such as MSE, chemical engineering, and chemistry who work in areas closely related to corrosion whose graduates possess the fundamental knowledge that would allow them to be quickly converted into corrosion experts. The committee estimates that a small percent of the graduate students advised by such faculty end up with careers as corrosion specialists. There are approximately 120 chemical engineering programs with graduate programs. Within these chemical engineering departments it is estimated that anywhere from 1 out of 7 to 1 out of 10 professors teach and or conduct research in areas closely related to electrochemistry or other subject matter related to corrosion. A typical graduate chemical engineering department has between 8 and 15 faculty members, 75 percent of whom carry out research. Assuming that each of these "research-active" professors specializing in a corrosion-related area produces 0.5-0.75 graduate student per year and that 10-25 percent of them find work as corrosion specialists, the annual supply of corrosion specialists from this route is 14.

Similarly, the 160 graduate mechanical engineering departments average 10-20 faculty members, with about 75 percent of them being active in research. In the best case, 1 in 20 of those professors in a mechanical engineering department might specialize in a field involving time-dependent materials properties, such as environmental degradation. Therefore, this country has about 60 mechanical engineering faculty capable of producing corrosion specialists. If each of these 60 professors graduates 0.5-0.75 student per year and 10 percent find jobs as corrosion specialists, 5 more people can be added to the supply of corrosion specialists. Similar estimates could be made for civil or nuclear engineering, but probably not for electrical, computer science, and systems engineering. Therefore, the total annual production of specialists and experts in corrosion is about 24-37 from programs emphasizing corrosion and another 20 or so from other engineering programs.

During the course of this study the committee learned that many people believe the number of corrosion faculty in the United States is declining. Among the 31 universities responding to the committee's questionnaire, 15 of the 26 answering the question said they would consider hiring a faculty member whose technical focus is corrosion. Of the remaining departments, only 2 (of 26) cited insufficient research funds as the deciding factor in not hiring such a faculty member; the rest said that other topics had higher priority. However, only 12.5 percent of those who would consider hiring new corrosion faculty would fill a newly created slot with requisite facilities set aside for the hiring of a corrosion specialist professor. Other answers were these: We would consider such a candidate if the candidate is competitive with candidates from other specialties. We do not have a specific position set aside for corrosion studies. Candidate must have strong materials expertise, not just corrosion. They will be considered if their work also involves applications of electrochemistry to energy production. Of those who would not consider hiring such a faculty member, 91 percent said that other topics have higher priority and 9 percent said that limited availability of research funds was the reason this area would not be considered. Other evidence of this trend came from a panel of university MSE department chairs interviewed by the committee. Most revealed that retiring faculty specializing in corrosion probably would not be replaced by younger colleagues. The committee's consensus was that grants for corrosion-related research were on the decline at these leading engineering universities. In addition, the universities recognize that more funding is available for novel and cutting-edge research. Acquiring such funding would allow them to compete for the best students and would satisfy the faculty desire to conduct research in new areas where important advances can be made.

The committee does not know how many corrosion specialists who retire in the next 10 years will be replaced. There was anecdotal evidence that faculty in corrosion are sometimes not replaced when a position is vacated. However, sometimes the replacement was a new person with competence in related areas such as electrochemical materials synthesis or fuel cells. The second piece of evidence for this perceived decline in faculty numbers is the shrinking number of journal articles on corrosion by authors at U.S. universities. Figure 2-7 shows data on the U.S. share of papers on corrosion and Figure 2-8 shows the U.S. share in two leading materials journals (more detail on the corrosion authorship data is available in Appendix C). Figure 2-9 shows a decrease in the number of Defense Technical Information Center reports on corrosion over the last few decades. Assuming that the mean authorship rate did not change with time, the number of faculty authors is decreasing with time.

DSB's report on corrosion control[10] points out that there is some fragmentation in corrosion funding with 1- and 2-year award periods. As a result of this fragmented funding, there is not enough continuity or stability to sustain gradu-

---

[10]Defense Science Board, *Report on Corrosion Control*. Available at http://handle.dtic.mil/100.2/ADA428767. Accessed March 2008. Summarized in Appendix A.

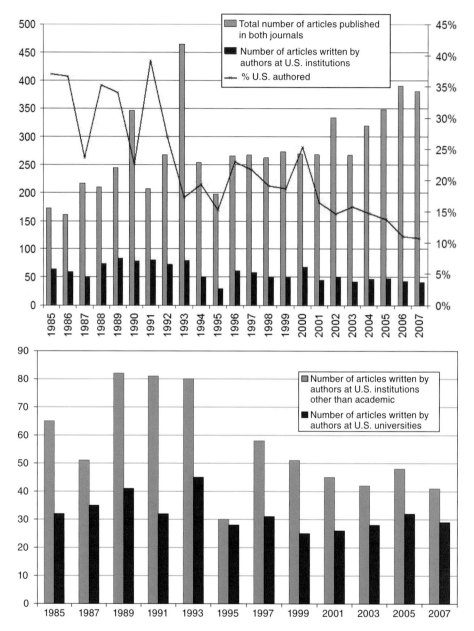

FIGURE 2-7 *Upper:* Articles published in *Corrosion* and *Corrosion Science* from 1985 to 2007. The chart indicates a gradual overall decline in the percentage of articles written by authors at U.S. institutions. *Lower:* Number of articles in the journals *Corrosion* and *Corrosion Science* written by authors at U.S. universities vs. by authors at U.S. institutions other than academic, tracked every other year between 1985 and 2007. Articles on property degradation in nonmetals invariably appear in specialized journals in polymer science, composites, ceramics, and so on.

FIGURE 2-8 Share of papers from the *Journal of Materials Research* and *Journal of Materials Science* written by authors with U.S. affiliations.

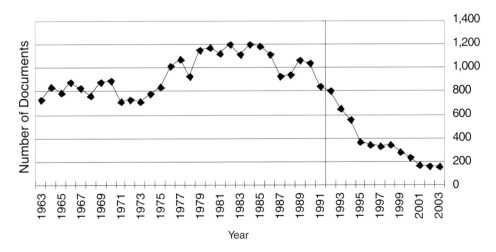

FIGURE 2-9 Number of documents on corrosion entered into the Defense Technical Information Center (DTIC) database. SOURCE: Advanced Materials and Processes Technology Information Analysis Center (AMPTIAC).

ate education over the 2- to 5-year time frame necessary for students to earn their degrees. These funding trends will also lead to a decline in the supply of corrosion specialists as well as in the number of papers on research in corrosion.

Since graduate schooling is the leading way to educate corrosion specialists and since graduate work is funded by research grants, it is reasonable to suppose that the supply of corrosion specialists is almost directly proportional to the number of grants and the total dollar value just as it is dependent on the number of faculty conducting corrosion research. To attract graduate students who eventually become corrosion specialists, university engineering programs offer graduate research assistantships (GRAs). Universities must have the financial resources to offer a GRA, which involves a research stipend, tuition, and health insurance, not to mention indirect costs of roughly 50 percent. R&D programs at universities and the funding of GRAs rely heavily on research grants and/or contracts. Much of this funding comes from the federal government and various state government agencies; some comes from industry and private foundations. Partial funding has recently been offered by technical societies.

How much does it cost to produce a corrosion expert through graduate education? The national average for funding a faculty member specializing in corrosion is $200,000, with wide variation. This amount supports between two and four graduate students or two postdoctoral research associates, assuming annual costs of $50,000 per student plus associated experimental and equipment costs as well as faculty time, raising the yearly costs to $80,000 to $100,000 per student. A master's-level student takes 2 years to complete the program, while a Ph.D. student takes 4-5 years. Overhead costs are about a third of the total. So the costs of educating a corrosion specialist are about $200,000 for a master's-level corrosion expert ($80,000 to $100,000 per year for 2 years) and $320,000 to $500,000 for a Ph.D.-level corrosion expert ($80,000 to $100,000 per year for 4-5 years). These numbers can be used to estimate how many new corrosion experts can be created for a given increment of research funding.

CONTINUING CORROSION EDUCATION

Those engineers in the upper part of the corrosion workforce pyramid (Figure 2-1), specialists who hold M.S. or Ph.D. degrees in corrosion, typically do not need to undertake continuing education except perhaps to learn a new technique or refresh their knowledge in it or an area they do not typically use, such as when moving to a new employer in a different technical area.

Engineers in the midsections of the pyramid—that is, those with a baccalaureate in engineering but without significant corrosion knowledge—can learn about corrosion in the workplace by means of employer-sponsored short courses that teach technical skills or basic knowledge. Few employers hire corrosion experts.

Rather, when they face complex corrosion problems they contract with technology consultants, who in turn employ corrosion specialists. This trend occurs with even greater frequency in smaller or medium-sized organizations, where employees are valued for their ability to perform many multidisciplinary tasks. Such employers often hire capable bachelor's-level engineers who can gain further competence through on-the-job training in corrosion design, mitigation, prevention, and control, supplemented by continuing education as time and resources permit.

This approach can help a company or government organization produce its own cadre of corrosion-knowledgeable engineers (and in some instances with further and more extensive knowledge-based education can even produce corrosion specialists). Another approach to enhancing corrosion knowledge could involve extramural or internal short courses. The former would involve additional living and travel expenses for someone attending classes at remote sites. Internal learning opportunities offered in nominally 1- or 2-hour segments during the normal workweek can include in-house distance or online learning, with employees still fulfilling most of their job functions.

It would make sense to tap employees who seem predisposed, by virtue of their technical background (in, for example, chemistry), competence, or general technical promise, to be able to acquire corrosion expertise (see Box 2-6). The continuing education in this instance would typically include from 3 to 10 years of experiential training combined with short courses and, possibly, stays at universities. One downside of this approach is that although these employees would emerge with good skills, they would probably have some gaps in fundamental knowledge compared to traditionally educated university graduates. This approach nonetheless would be superior to leaving the organization with no corrosion expertise. Clearly the duration of the on-the-job training cycle depends primarily on the

---

BOX 2-6
Cost of Producing a Corrosion Expert by
Means of Continuing Education

It would be useful to estimate an organization's total cost for producing a corrosion expert in this way. It can take from 3 to 5 years for a new, B.S.-level engineer to be fully productive, with much of this time needed for learning operating processes and applications. While most new employees will require additional formal education, all must receive on-the-job training. Assuming that (1) the typical burdened annual labor cost for an engineer is $200,000 per year and (2) the process of learning about corrosion requires up to 50 percent of that engineer's time over 5 years on the job before that engineer can be declared a corrosion specialist, the prorated cost to that organization would be $600,000 per expert (50 percent × $200,000 per year × 5 yr = $500,000, with another $100,000 for tuition and off-site travel = $600,000).

individual's capability and the availability of experienced, qualified mentors as well as the scientific or technical complexity of the job assignments. An organization might consider sponsoring an employee as a full- or part-time student to earn an advanced academic degree.

The corrosion workforce pyramid (Figure 2-1) shows that the foundation of the ideal corrosion workforce is a team of corrosion-aware and corrosion-knowledgeable technologists. This segment of the workforce includes maintainers, technologists, and some procurement, production, and maintenance officials who require only minimal corrosion knowledge. For this segment, continuing education in the form of informal on-the-job training and formal short courses is a cost-effective way to develop more productive and competent employees. In particular it can be useful for workers who perform routine and repetitive duties.

Another reason for offering courses to technologists is for their qualification and certification. A certificate from a knowledgeable, independent third party proves that a worker is qualified to perform a particular procedure (e.g., surface preparation, application of coatings and/or linings, cathodic protection) or to inspect systems or components for the integrity of painted, coated, or lined systems, for example. Job performance in this sector of the workforce can be enhanced when these employees have a better understanding of corrosion's impact. They will be able to identify corrosion and proactively prevent it from degrading the performance and durability of the particular system or piece of equipment they are maintaining.

Because short courses are an essential element of continuing education, the committee carried out a search for such courses. Appendix D provides an overview of the short courses that the committee is aware of (there may be others of which the committee is not aware). Short courses cost from $100 to $3,800 depending on their length, the means of instruction, and the organization providing the instruction. Table 2-1 summarizes the material in Appendix D organized roughly according to the categories in the corrosion workforce pyramid.

TABLE 2-1 Courses Organized into Basic or Advanced Generic Corrosion, System- or Technology-Specific Groupings Plus Additional Certification or Training Courses

Category	Course Provider	Structure	Focus
Advanced (post-B.S.) courses on generic corrosion	Penn State	Short course, with lab	General
	North Dakota State	Short course, with lab	General
	ASM	Short course, without lab	General
	Society of Automotive Engineers	Short course, without lab	General/automotive
	Ohio State University	Distance learning	General

continues

TABLE 2-1 Continued

Category	Course Provider	Structure	Focus
Introductory (undergraduate) courses on generic corrosion	NACE	Short course, with lab	General
	Corrosion College	Short course, with lab	General
	Technology Training, Inc.	Short course without lab	General
	Western States Corrosion Seminar	Short course without lab	General
Applied courses on platform- or system-specific corrosion	University of Kansas	Short course with lab	Aircraft
	NACE	Short course with lab	Pipeline
		Short course with lab	Refining industry
		Short course with lab	Offshore
		Short course with lab	Shipboard
	Oklahoma State University	Short course without lab	Pipeline
		Short course without lab	Internal
	Appalachian Underground	Short course without lab	Pipeline (basic)
		Short course without lab	Pipeline (advanced)
		Short course without lab	Water and wastewater
		Short course without lab	Coatings
	Corrosion Clinic	Short course without lab	Defense industry
		Short course without lab	Automotive industry
	Center for Professional Advancement	Short course without lab	Oil and gas industry
	Corrosion Courses	Short course without lab	Oil and gas industry
	PetroSkills, LLC	Short course without lab	Oil and gas industry
Courses on applied technology-specific corrosion	Oklahoma State University	Short course without lab	External
	Purdue Underground Short Course	Short course without lab	Cathodic protection (basic)
		Short course without lab	Cathodic protection (advanced)
		Short course without lab	Coatings
	Technology Training, Inc.	Short course without lab	Corrosion control techniques
	Western States Corrosion Seminar	Short course without lab	Corrosion fundamentals
		Short course without lab	Intermediate-level corrosion
		Short course without lab	Advanced-level corrosion
Supplemental corrosion awareness	Defense Acquisition University	Online	Corrosion prevention and control
	Army Corrosion Training (CTC)	Online	Basic corrosion control course
Training to obtain a certificate or license	NACE	Short course with lab	Coatings Inspector (1)
		Short course with lab	Coatings Inspector (2)
		Short course with lab	Cathodic protection

NOTE: ASM, American Society for Materials International; CTC, Concurrent Technologies Corporation.

Short courses are primarily taught by three types of organizations: professional societies, postsecondary institutions, and private companies. Appendix D lists the organizations that are now regularly offering extramural corrosion training and education, along with some of the courses and administrative and course content details. Topics range from the basic and fundamental—as might be expected, a course entitled Corrosion Basics—to the focused and specific, such as one entitled Corrosion in Microelectronics. Corrosion is also covered in some short courses in the context of overall component design—for example, the ASME course Mechanical Insulation Design.

All the courses cover the following:

- Introductory courses giving an overview of corrosion and its importance for society;
- The mechanisms of corrosion, including electrochemical, pitting, and cracking and its thermodynamic and kinetic aspects;
- Materials-specific classes covering both materials selection and how corrosion mechanisms vary between materials;
- Corrosion control by cathodic and anodic means and by coatings;
- The detection of corrosion; and
- Sector-specific courses in sectors such the military, pipelines, the automotive industry, and aircraft applications.

Table 2-2 summarizes the committee's analysis of how different levels of the workforce would benefit from the courses available.

SUMMARY OF FINDINGS

The committee has found that corrosion technologists are often trained on the job by means of short courses focused on defined sets of skills and on responses to generally known sets of conditions that are often repeated over and over again. It has also found that only a fraction of U.S. undergraduate MSE students are exposed to a course with detailed information on corrosion. The availability of such a course depends on faculty interest and expertise, as well as on how the teaching of corrosion fares in competition with other demands on the curriculum. In other design and engineering disciplines, undergraduate engineering students typically learn little about materials selection and usually have no more than one or two lectures on corrosion, often none. Whereas graduate engineering students specializing in corrosion get formal training in it, graduate MSE students are typically not required to take a course in corrosion; moreover, such courses are only offered in departments where there is a faculty member with expertise in corrosion. The availability of teachers for corrosion depends in turn on the health

TABLE 2-2 How Continuing Education Courses Like Those Listed in Appendix D Would Benefit the Corrosion Workforce

Level of Corrosion Expertise	Level of Education and/or Training	Training Options and Outcomes	
		Short Courses	Online Courses
Expert	Ph.D. corrosion faculty	n/a	n/a
	Ph.D.	Would be a refresher	n/a
	M.S. with extensive on-the-job training	Would supplement knowledge	n/a
Highly knowledgeable	M.S. with on-the-job training	Would supplement knowledge	n/a
	B.S. in corrosion field with extensive on-the-job training	Would supplement knowledge	n/a
Knowledgeable	Ph.D. in related field with no specific corrosion experience	Would increase knowledge	Would increase knowledge
	B.S. with on-the-job training	Would supplement knowledge	Would increase knowledge
	B.S. in noncorrosion field with extensive on-the-job training	Would increase knowledge	Would increase knowledge
Minimally knowledgeable	B.S. in noncorrosion field with minimal on-the-job training	n/a	Could increase corrosion awareness and job performance
	No degree but on-the-job training	n/a	Could increase corrosion awareness and job performance
Minimal to none: Familiarity would be beneficial	Procurement official with nontechnical degree	n/a	Could increase corrosion awareness and job performance
	Maintenance and production worker with trade school education	n/a	Could increase corrosion awareness and job effectiveness
	Maintenance and production worker with high school education	n/a	Could increase corrosion awareness and job effectiveness
	Maintenance technicians and military personnel	n/a	Could increase corrosion awareness and job effectiveness

NOTE: n/a means the committee believes the course would not benefit that category of worker relative to its effort or cost.

of the corrosion research community and therefore on the availability of funding for that research.

The committee has found that there are many short courses available for the continuing education of engineers and technologists of many different skill and education levels in addition to on-the-job training programs (see Appendix D). Since continuing education often imparts specific skills in specific technologies during intensive, usually 2- to 5-day extramural courses, it often leaves gaps in the employee's fundamental knowledge base (compared with the traditionally educated university graduate who takes semester-long courses where course prerequisites and out-of-class assignments assure a better learning of corrosion fundamentals). In the committee's opinion, anything learned from short courses, while beneficial, is not as deep as the learning from a rigorous corrosion education curriculum that teaches basic science, engineering, and mathematics and gives an engineer the intellectual skills to perform complex tasks, create new materials and innovative processes, and solve difficult problems that enable the control and mitigation of corrosion.

3

Conclusions and a Recommended Path Forward

Chapter 1 discussed how the corrosion of materials, leading to the degradation of their physical properties, is of great concern to society. Chapter 2 focused on the current state of corrosion education and its general shortcomings. The content of those chapters was based on the information elicited from the academic sector by a Web-based questionnaire, information shared at the two town meetings held at professional society meetings, information gathered informally between meetings, and from information and opinions gathered at the committee's meetings. This chapter draws some conclusions from the findings in those two chapters, from the information gathered by the committee at the 2007 National Academies Materials Forum,[1] and from the government, industry, and academic panels the committee convened at its meetings during the course of the study (see Appendix E for the relevant agendas). It discusses the impact of the current situation on the education of the nation's engineers, on the government and its assets, and on industry and its interests. In this chapter the committee analyzes the degree to which the existing education system has equipped the workforce at all levels to mitigate and minimize corrosion and assesses whether this education is adequate, whether current educational trends are going in the right direction, and whether a different path is needed. It concludes by recommending a path forward, with specific actions recommended for government, industry, academia, and the corrosion science and engineering community.

---

[1]National Research Council, *Proceedings of the Materials Forum 2007: Corrosion Education for the 21st Century*, Washington, D.C.: National Academies Press (2007). Available at http://books.nap.edu/catalog.php?record_id=11948. Accessed January 2008.

THE IMPORTANCE OF CORROSION EDUCATION

Corrosion has been studied by scientists and engineers for about 150 years and remains relevant in almost every aspect of materials usage. As was demonstrated in Chapter 1, corrosion can have a great impact on the safety and reliability of an extremely wide range of articles of commerce, and its financial impact in the United States is very large. Examples of technology areas where corrosion plays an important role include energy production (for example, power plant operation and oil and gas exploration, production, and distribution), transportation (for example, automotive and aerospace applications), biomedical engineering (for example, implants), water distribution and sewerage, electronics (for example, chip wiring and magnetic storage), and nanotechnology. It is reasonable to consider that the increasingly harsh physical environments to which critical systems such as energy production are subjected (one example is nuclear reactors that operate at high temperatures) and the proliferation of new technologies in support of societal goals (for example, the growing use of hydrogen as an auto fuel) may increase the cost of corrosion to society unless mitigating steps are taken. It has been estimated that remedial actions based on a better and more widespread understanding of the corrosion phenomenon could reduce significantly the financial burden of corrosion to the nation. Although insufficient corrosion education in the engineering profession is not the only reason for the absence of such actions, the committee has concluded that it is a major one.

Successful application of corrosion knowledge and understanding could save billions of dollars annually. Teaching engineers the fundamentals of corrosion and corrosion prevention is critical to both mitigating the damage done by corrosion as well as to the competitiveness of the nation's industries and the effectiveness of its defense. The automotive industry is one example of the value of corrosion awareness in design. The use of galvanized steel body panels and improved painting methods have improved the durability of car exteriors in relation to corrosion. However, the need to save weight has led the automotive industry to consider extensive use of magnesium. This is a paradigm shift that will require extensive advances in corrosion knowledge on the part of manufacturers, their suppliers, and maintenance organizations. An engineering workforce that does not know enough about corrosion will have a difficult time dealing with such paradigm shifts in particular and corrosion problems in general.

CONCLUSION 1. Corrosion, or the degradation of a material's properties as a result of its interaction with the operating environment, plays a critical role in determining the life-cycle performance, safety, and cost of engineered products and systems.

CONCLUSION 2. Advances in corrosion control are integral to the development of better technologies that make current, legacy, and future engineered products, systems, and infrastructures more sustainable and less vulnerable. Such advances will require corrosion-knowledgeable engineers and an active corrosion research community.

CONSEQUENCES OF THE CURRENT STATE OF CORROSION EDUCATION

As discussed earlier in this report, most curricula in engineering design disciplines require engineers to take a course in materials engineering, which typically covers some basics of the relationships between structure, properties, and processing.[2] While such a course would make an engineer aware of issues related to materials selection, corrosion, if covered at all, is usually discussed in only one lecture at the end of the course. The concepts related to materials selection in general and corrosion specifically are usually not reinforced in the other parts of the curriculum. As a result, graduating engineers have little understanding of corrosion in metals or how to design against it and even less when it comes to the degradation of nonmetals.

Even those with bachelor's degrees in materials science and engineering (MSE) or related fields such as metallurgy, ceramics engineering, and so on receive little or no education in corrosion science and engineering. Because there is significant pressure on MSE departments to include emerging areas such as nanotechnology and biomaterials, corrosion and other longer established areas of materials engineering are losing out.

The committee is convinced that advances in the durability and longevity of engineered materials and the savings that will accrue are more likely if engineers understand the fundamental principles of corrosion science and engineering and apply them using best engineering practices. This conviction is based on great opportunities in three areas:[3]

---

[2]ABET, the recognized accreditor for college and university programs in applied science, computing, engineering, and technology, defines engineering design as the process of devising a system, component, or process to meet desired needs. It is a decision-making process (often iterative) in which the basic science and mathematics and engineering sciences are applied to convert resources optimally to meet a stated objective. Engineering design disciplines include mechanical engineering, civil engineering, aeronautical and aerospace engineering, and so on.

[3]The *Corrosion Costs* study, carried out in 2001 for the FWHA and NACE International, noted that technological changes have provided many new ways to prevent corrosion and put available corrosion management techniques to better use. However, better corrosion management can also be achieved using technical and nontechnical preventive strategies. For a summary from NACE International, see http://events.nace.org/publicaffairs/images_cocorr/ccsupp.pdf. Accessed October 2008.

- Design practices for better corrosion management;
- Life prediction and performance assessment methods; and
- Improved corrosion technology through research, development, and implementation.

Degradation of materials must be anticipated and minimized as much as possible by the proper design, use, and maintenance of materials. Strategies for making technological advances and the development of best practices in the management of materials will depend on

- Understanding current design practices for corrosion control;
- Utilizing methods for predicting materials life and performance;
- Exploiting advanced technologies for the research, development, and implementation of new and better corrosion-resistant systems; and
- Developing strategies for realizing savings.

The ability of the nation's technology base to develop these methodologies and technologies depends on an engineering workforce that understands the physical and chemical bases for corrosion as well as the engineering issues surrounding corrosion and corrosion abatement. Consider the role of engineering in bridge design and construction. One would not design a bridge without considering fatigue loadings. Nor should it be designed without considering the continuous degradation of its materials by the environment in which it operates.

Both the public and private sectors appreciate the need for engineers who have been taught corrosion engineering so that they can take corrosion into account during design and manufacture. The importance of corrosion education in today's world continues to increase as the limits of material behavior are stretched to improve the performance of engineered structures and devices. Employers recognize the need for employees with competence in corrosion engineering, but as this report reveals they are not finding it in today's graduating engineers, who have no fundamental knowledge of corrosion engineering and little understanding of the importance of corrosion in engineering design and do not know how to control corrosion in the field. In fact, the problem has become so critical that a principal concern of employers is that those making design decisions don't know what they don't know about corrosion. At the very least, it would benefit employers if they required that all engineers making design and materials selection decisions (see Box 1-3) at least know enough about corrosion to understand when to bring in an expert.

For the purposes of this report, and as suggested in the charge to the committee, the workforce has been divided into graduating engineers and practicing engineers. The committee found it helpful to conduct its assessment of practicing engineers in terms of the impact of the current system on two sectors: government and industry.

Graduating Engineers

As discussed in Chapter 2 in relation to the so-called corrosion workforce pyramid, the corrosion workforce can be divided into a number of categories relevant to this report.

1. Technologists needed to perform repeated crucial functions, such as paint inspectors and specifiers, and cathodic protection designers and installers.
2. Undergraduate engineering students in MSE who upon graduation should be knowledgeable in materials selection;
3. Undergraduate engineering students in other design and engineering disciplines such as mechanical, civil, chemical, industrial, and aeronautical engineering; and
4. MSE graduate students who upon graduation should be very knowledgeable in materials selection and in some cases will go on to be experts in the field of corrosion.

The committee has found that corrosion technologists are often trained through the supervised performance of repeated and predictable corrosion tasks (on-the-job training), in conjunction with short courses and associate degrees offered by a few community colleges. The tasks performed by these corrosion technologists often require implementation of standardized practices. This training generally equips an individual to recognize a fairly well-behaved set of conditions and teaches how he or she would go about selecting the preferred solution. However it does not impart enough understanding so the individual could apply a body of knowledge to a situation he or she had not encountered before.

The committee has found that at only a fraction of the MSE departments across the country do undergraduate MSE students take a course with some detailed corrosion content. The availability of such a course depends on faculty interest and expertise and how well corrosion competes with other subjects demanding a slot in the curriculum. In other design and engineering disciplines, undergraduate engineering students typically take one course, a survey course, in materials. But they learn little about materials selection and usually would have attended no more than one or two lectures on corrosion, if that.

Whereas graduate engineering students specializing in corrosion get formal training in corrosion, graduate MSE students are typically not required to study it, and a corrosion course is offered only in departments where a faculty member has expertise in corrosion. The drop in U.S. publishing in corrosion science and engineering (see Figure 2-7) indicates that research activity in corrosion has declined. The committee speculates, although with some confidence given the consistent anecdotal evidence it received from several quarters, that the decrease in publishing is concomitant with a decrease in the number of faculty with such expertise and, by extension, in the number of those who could teach the subject.

It seems this situation is set to continue. According to the evidence the committee heard, many of the highest ranked and most prestigious MSE departments in the country have no interest in creating or maintaining a corrosion research program. If taught at all in such departments, corrosion would be taught either by a faculty member with no intimate knowledge of the field or by someone with expertise in a related area, such as batteries or fuel cells. The committee recognizes that the inclusion of new course material—both required and elective—in engineering curricula makes it difficult to also cover topics like materials selection in general and corrosion in particular.

CONCLUSION 3. Corrosion engineering education is not a required element of the curriculum of most bachelor's-level programs in materials science and engineering and related programs. In many programs, corrosion engineering education is not offered. As a result, most engineers graduating from bachelor's-level materials-related programs have an inadequate background in corrosion.

CONCLUSION 4. The bachelor's-level education of engineers who serve on design teams involves too little detail in corrosion-relevant materials selection and almost no exposure to corrosion education in general. This lack of knowledge and awareness ultimately jeopardizes the health, wealth, and security of the country.

CONCLUSION 5. Undergraduate and graduate education in the field of corrosion engineering requires an adequately funded university research community.

Practicing Engineers in Government and Industry

The lack of exposure to corrosion engineering principles and practices in their educational experience is a serious flaw in the training of many practicing materials engineers and design engineers. It appears to the committee that government agencies are particularly lacking in in-house corrosion experts. This is partly because such agencies believe they can outsource the search for the solution of a corrosion problem to external consultants and partly because they feel they cannot find corrosion experts to hire. For many of the same reasons, industry often ignores corrosion until a major problem occurs. Smaller companies tend to rely on vendor information. Most companies have few corrosion experts. They prefer to hire people with broad rather than specialized backgrounds and provide in-house training in corrosion. If there is no in-house experience, companies will outsource problems to consultants or to the vendors. As trained corrosion engineers retire, the committee is concerned there will be a shortage of trained people to hire as replacements.

The implementation of effective corrosion prevention strategies requires an educated workforce of practicing engineers. In the context of this report's consideration of the current state of affairs in corrosion education, it is important to understand the needs of the government agencies—the Department of Defense (DOD), the Department of Transportation (DOT), the U.S. Army Corps of Engineers, and others—whether those needs are being met, and, if not, how the gaps in workforce understanding can be addressed.

Current Workforce

What does the community of practicing corrosion engineers look like? According to a survey of the U.S. membership of NACE International, a majority of self-identified corrosion engineers have a background in mechanical engineering, chemical engineering, or materials science (Figure 3-1). A NACE International survey of its membership (Figure 3-2) shows that most members function as engineers, managers, technologists, sales and marketing professionals, research scientists, or consultants. Only 34 percent have more than a bachelor's degree. More than one-half (54 percent) of corrosion protection practitioners have not taken a course in corrosion during their formal education. A large number (45 percent) began employment at the technician level before moving into the field of corrosion control. A large number (44 percent) of the active practitioners plan to retire or move to another field in the next 10 years. Close to one-half (48 percent) of the respondents think their position will be filled by someone with similar credentials and experience, and a large number (42 percent) said that their companies require a 4-year degree for the position they are currently occupying. Three-quarters of the respondents were between 41 and 65 years old, with 41 percent between the ages of 51 and 65 (Figure 3-3). Over half the current workforce has no formal education in corrosion and 90 percent of respondents think the corrosion education of current graduates is fair or poor (Figure 3-4).[4]

---

[4]The University of Akron survey of employers found that three-quarters of respondents saw "a shortage of qualified job candidates with corrosion engineering skill sets." In the same survey, two-thirds of respondents thought that engineering graduates are "not equipped with an acceptable level of understanding when it comes to the effects and management of corrosion."

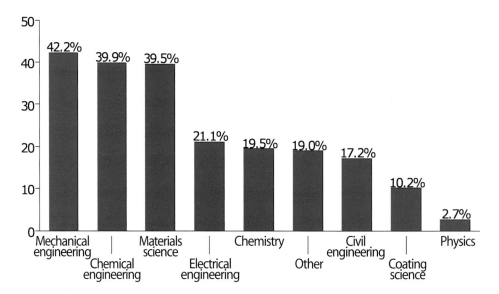

FIGURE 3-1 Make-up of the corrosion community by field. The survey asked for the educational specialization of staff hired for corrosion engineering positions (more than one answer was allowed, so the total exceeds 100 percent). SOURCE: Copyright 2008 Eduventures, Inc. Copyright 2008 the University of Akron. All rights reserved. Research conducted by Eduventures under contract for the University of Akron.

As to whether there is continuing demand for corrosion graduates, a recent article and some information gathered by the committee indicate that the demand for corrosion professionals remains strong. The committee did an on-line search for engineer jobs on two popular career Web sites (Table 3-1). The search showed that there is a healthy demand for corrosion professionals. More compelling data were gathered for a recent report published in *Materials Performance*.[5] The article reports that the NACE career center received 168 job postings between January 1, 2007, and October 24, 2007, up from 162 job postings in the whole of 2006. Corrosion positions in the engineering category accounted for 30 percent of the job postings, followed by technician (20 percent), inspector (10 percent), management (8 percent), research (8 percent), sales/marketing (5 percent), and "all other categories" (19 percent). About 28 percent of the jobs posted were located in Texas. CorrosionJobs.com received between 75 and 100 job listings annually, with corrosion technicians being the most sought after on that site, followed by specialists, engineers, project managers, and researchers. About half of the listings on that site

[5]Kathy Riggs Larsen, "Wanted: Corrosion Professionals," *Materials Performance*, December 2007.

Primary Job Function

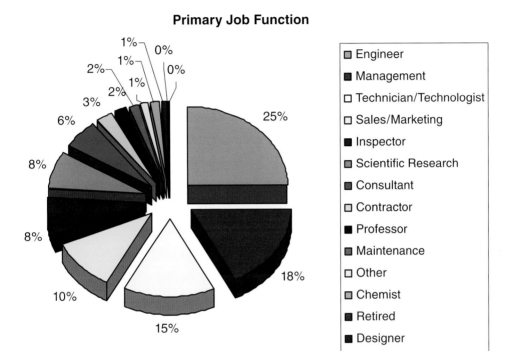

Education Level

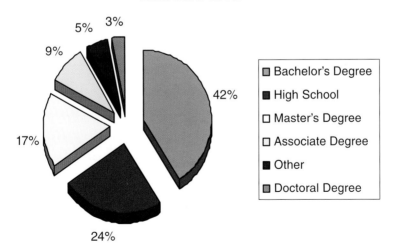

FIGURE 3-2 Results of NACE survey of its membership. SOURCE: Aziz Asphahani and Helena Seelinger, NACE Foundation, "The Need for Corrosion Education," Presentation at the Materials Forum 2007: Corrosion Education for the 21st Century.

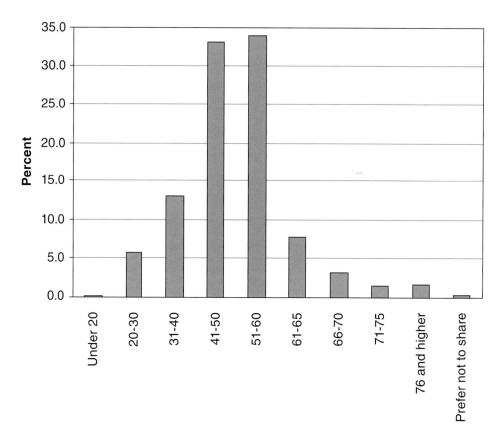

FIGURE 3-3 Age distribution of NACE International membership. The number of respondents was 1,595. SOURCE: Aziz Asphahani and Helena Seelinger, NACE Foundation, "The Need for Corrosion Education," Presentation at the Materials Forum 2007: Corrosion Education for the 21st Century.

come from service companies, 40 percent from pipeline and operating companies, and roughly 5 percent from state transportation departments. Service companies in Houston, Texas, accounted for 30-40 percent of the employers.

Government

The committee heard from a panel of government representatives (the government agencies and their representatives, along with other agenda details of the committee's meetings, are listed in Appendix E). Based on these discussions, private informal data gathering by committee members during the course of the study, and the committee's own experience, there are a number of important findings.

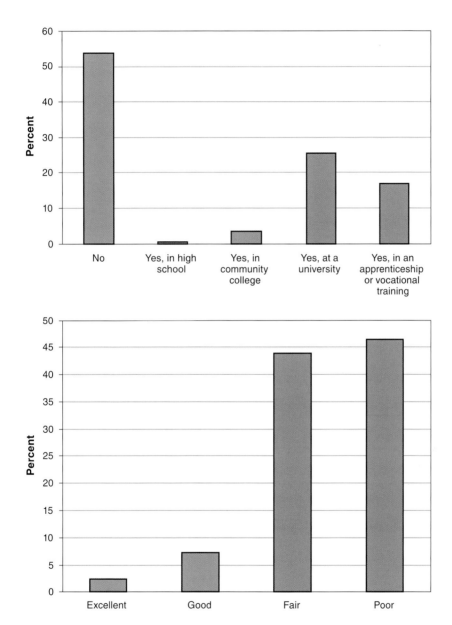

FIGURE 3-4 *Upper:* Educational background of current corrosion workforce. The question asked was, Did you take "corrosion courses" in your formal education, including courses on corrosion prevention technologies? (check all that apply). The number of respondents was 2,396. *Lower:* Opinions on the education of recent graduates. The question was, When hiring recent engineering graduates, how would you rate their knowledge of corrosion-related topics? The number of respondents was 41. SOURCE: Aziz Asphahani and Helena Seelinger, NACE Foundation, "The Need for Corrosion Education," Presentation at the Materials Forum 2007: Corrosion Education for the 21st Century.

TABLE 3-1 Demand for Corrosion Professionals

Search term	HotJobs April 23, 2008	Monster.com May 12, 2008
Materials engineer	47	56
Materials scientist	5	13
Polymer engineer	11	4
Corrosion engineer	12	11
Corrosion scientist	1	2
Biomedical engineer	108	51
Metallurgical engineer	26	28
Mining engineer	16	19

NOTE: As measured by the number of positions advertised on two online employment Web sites. To get an idea of the demand for corrosion experts in today's job market, searches were performed on HotJobs.com and Monster.com for corrosion engineers and scientists, along with other types of engineers. With the exception of "biomedical engineer," the numbers from the two sites were remarkably similar.

Finding. There is often no in-house corrosion expertise in the agencies represented on the panel.[6]

The overwhelming sense of the discussion with the government panel led the committee to conclude that maintaining a corps of in-house corrosion experts is not now and probably never was a high priority. Likewise, the committee sensed the panel's belief that although it is a good idea for engineers to have some corrosion knowledge, current management appears to expect project managers to find an expert on demand when projects require that expertise, largely by outsourcing the job to a contractor or consultant. This situation is aggravated by the projected retirement of the few people with corrosion expertise and the absence of corrosion engineering experience in new hires. Despite the importance of materials selection in the design of any new engineering system, materials engineers typically make up only a small fraction of the engineering workforce. Corrosion engineers are an even smaller fraction. As a consequence, corrosion is often the last issue considered in design and is not treated with good fundamentals or good technology.

Finding. Younger engineers working for the government have had little if any exposure to classes on corrosion and are unaware of the consequences of corrosion.

New hires tend to come from disciplines such as mechanical engineering, civil engineering, electrical engineering, and chemical engineering. Few if any of these

---

[6]Government agencies represented on the panel were the Hanford Nuclear Waste Storage Facility, the Army Aviation and Missile Command, the U.S. Army Corps of Engineers, the U.S. Bureau of Reclamation, and the Virginia Transportation Research Council.

new hires will have had any exposure to corrosion engineering instruction during their university studies. As such, their ability to diagnose corrosion problems and to introduce successful remediation strategies is limited.

> **Finding.** Since corrosion engineers can be hard to find and to hire, retiring corrosion engineers in government are not being replaced, and much of what is needed in terms of corrosion engineering knowledge is being learned on the job by other engineers.

The committee heard from its panelists, backed up by the experience of those on the committee who are in the business of teaching engineers about corrosion, that demand on the part of industry for the few corrosion experts who are being graduated means that government agencies are having a hard time competing for the expertise. Moreover, the fact that government primarily hires U.S. citizens reduces the pool of eligible graduating engineers even further. In response, some agencies, like the U.S. Army Corps of Engineers, have developed internal corrosion action teams to provide short courses and in-house training for practicing engineers who have little or no corrosion education. While there is no doubt that experiential learning is valuable, some organizations interested in diagnosing a corrosion problem and remediating it will require workers who understand the fundamentals, which comes from formal education aimed at producing a corrosion expert. Though metallurgy is still taught at nine ABET-accredited universities in the United States, representatives of some agencies indicated to the committee that they no longer hire metallurgists. They also stated that they are hiring few materials engineers, so that as retirements proceed an essential part of the engineering workforce is being lost.

> **Finding.** Cost, readiness, safety, and reliability are key elements of corrosion control in defense systems. That notwithstanding, corrosion still struggles to get attention.

Remarkably, the government panel reported that the costs associated with corrosion are still not getting a lot of attention in many parts of the government. Indeed, it appears that on occasion corrosion mitigation is delayed in order to meet short-term financial goals.[7] This creates a gap in the in-house knowledge and

---

[7]The report *Corrosion Costs* notes that there is often a disconnect between those who control corrosion costs and those who incur the costs. This can lead to a mentality of "build it cheaper and fix it later" and a disregard for life-cycle costs. The situation is exacerbated when the builder is not made responsible for the repair costs (for example, federal funds are used to build bridges, yet state funds are used to maintain the bridges). This can lead to conflicts in the trade-off between lower construction costs and higher maintenance costs. In addition, the indirect costs of corrosion, often borne by the public, may not be allocated to the owner/operator. Conversely, the owner/operator cannot take credit for or receive additional compensation for long-term savings.

expertise available to the government. To deal with this gap the government has taken the following actions:

- Management looks to bring in outside experts in on a case-by-case basis through commercial outsourcing and consulting services.
- Staff attend short courses presented by professional societies such as ASM and NACE International.
- In-house training programs are often the solutions of choice—that is, people are hired and taught what they need to know.

Delivering good courses on site, either by professional societies or by university faculty, seems to be a key method of providing instruction and learning opportunities to staff who need to have in-depth knowledge of corrosion engineering or at least an awareness of some of its aspects.

Industry

The committee heard from a panel of industrial representatives and learned that the primary impact of the scarcity of corrosion engineers is on maintenance and plant operations. An exception is in the oil field and energy generation sectors, where corrosion control is receiving some attention in the design stages as a result of the demonstrated economic, environmental, and societal implications of corrosion failures in these systems. Maintenance and replacement costs are not the main drivers for concern in these sectors. Rather, the costs of product loss, safety lapses, and contamination of the surrounding environment following a system's failure are what is driving the concern. The committee also learned about another case where corrosion plays a large role and its control is included in the design stage: increasing the range of operating temperature for gas turbines, which improves their efficiency. This case is clearly related to the ability of turbine manufacturers to increase their competitiveness and, accordingly, their profit margin by marketing more efficient energy-generating machines.

However, more often than not, corrosion is not a priority for senior management in industry. It is only when dramatic failures attributable to corrosion occur that either senior management or government regulators become interested in corrosion-related design or corrosion control systems.

Given that the purpose of this study is to address the state of corrosion education in the United States, the committee attempted to summarize its finding on the current needs of industry based on the testimony it heard from the industrial panel it had assembled and on its own expertise.

Finding. Corrosion is generally treated as a maintenance and operational issue, not as a design issue.

Industry needs for corrosion engineering education generally parallel those expressed by government panelists. Specifically, industrial panelists bemoaned the fact that few if any trained corrosion specialists are available. This is especially true in plant and product design. While there are many instances where corrosion is included in design considerations (Box 3-1), in general it is not, with the result that unscheduled outages and extraordinarily expensive repairs often ensue. Corrosion is often treated as an aftermarket event, where inspection and sporadic maintenance, often unanticipated, are the rule. Exceptions are industries where corrosion damage can have near-catastrophic consequences. For example, in the oil industry, pipeline corrosion can lead to severe environmental contamination as well as loss of valuable product. Other industries where corrosion is typically (but not always) considered in design are where worker and public safety are involved. Examples are the energy production industry (notably the nuclear power industry), the aviation industry, and the chemical process industry. Still another reason for taking corrosion-related phenomena into account in the design is life-cycle considerations. Too often corrosion-related damage is relegated to the sphere of failure analysis, after failures occur. Seldom is equipment or a component replaced proactively to forestall corrosion. The reliability of in-plant equipment, as well as of consumer products, is also an issue. The cost of product recalls necessitated by inadequate attention to corrosion at the design stage is monumental, exceeding hundreds of millions of dollars in the automotive industry alone, where significant improvements in corrosion control have been made.[8]

Finding. Management, especially of small to medium-sized companies, often depends on vendors to supply materials specifications for corrosion control.

Large companies may have the luxury of having staff members who spend at least some their time dealing with corrosion issues. However, small to medium-sized companies generally do not have trained, or even trainable, engineers who are equipped to deal with simple corrosion-related tasks such as materials selection, periodic inspection, appropriate inspection, maintenance of corrosion protection programs, or even a general knowledge of materials/environment compatibility.

---

[8]The *Corrosion Costs* report notes that in the late 1970s, automobile manufacturers started to increase the corrosion resistance of vehicles by using corrosion-resistant materials, employing better manufacturing processes, and by designing more corrosion-resistant vehicles through corrosion engineering knowledge. Because of the steps taken by the manufacturers, today's automobiles have very little visible corrosion, and most vehicles fail mechanically before they wear out structurally. However, the total annual cost incurred is high, and much can be done to further reduce the cost. The total cost of corrosion to owners of motor vehicles is estimated at $23.4 billion per year.

BOX 3-1
Failure Mode and Effects Analysis

Designers should use the process of failure mode and effects analysis (FMEA) to identify the mode of failure in every component in a system. A failure is the inability of a system to meet a customer's requirement as opposed to actual material breakage or failure. The FMEA methodology predicts what failures could occur, predicts the effect of the failure on the functioning of the system, and identifies the methods that should be used to prevent the failure. For corrosion to be adequately included in the FMEA, designers need to know more about the mechanisms of corrosion failure than they currently do.

Accordingly, where corrosion-knowledgeable staff are not available or affordable, vendor information is often used for making decisions about new or replacement manufacturing equipment. Consultants are sometimes brought in, but more often to diagnose failures than to participate in the design.

Finding. The supply of materials engineers and corrosion engineers is growing tighter.

Some of the panelists perceived that engineering staffs, in general, are being reduced. Notwithstanding the accuracy of this perception, in recent years a constant (or possibly somewhat declining) number of materials engineers have been produced by American universities (Table 3-2 and Figure 3-5). Given that corrosion courses in the universities are often taught by faculty who are near retirement or already retired and that there seems to be little appetite for replacing them, the supply of materials engineers with knowledge of corrosion is likely to decrease further. Faculty who are replaced will be replaced by faculty who will be expected to perform functions other than teach corrosion.

To reiterate, there are too few engineers with solid training in corrosion science, technology, and control processes entering the job market. In many if not most cases, companies are training engineers from other disciplines to address specific corrosion problems. These engineers may lack a comprehensive understanding or a knowledge of the things that can be done to address the source of corrosion-related failures or may not be able to recommend materials and practices that will prevent future failures. Professionals trained in or cognizant of corrosion processes at elevated temperatures are almost nonexistent, and only one major university in the United States has an active research program in high-temperature corrosion.

CONCLUSION 6. DOD's recent proactive stance on corrosion control will be undermined by the shortage of engineers and technologists with sufficient corrosion engineering education.

TABLE 3-2 Materials and Metallurgical Engineering Degrees Awarded in the United States, by Degree Level and Gender of Recipient, 1966-2004

Academic Year Ending	Bachelor's			Master's			Doctorate		
	Total	Men	Women	Total	Men	Women	Total	Men	Women
1966	792	785	7	400	397	3	211	209	2
1967	836	828	8	444	443	1	267	266	1
1968	881	863	18	460	458	2	215	213	2
1969	952	942	10	441	435	6	280	279	1
1970	977	967	10	429	423	6	303	302	1
1971	916	903	13	480	472	8	306	305	1
1972	909	893	16	524	513	11	294	291	3
1973	885	870	15	582	569	13	299	292	7
1974	821	789	32	521	508	13	280	277	3
1975	711	676	35	500	483	17	272	267	5
1976	704	661	43	475	447	28	252	244	8
1977	738	679	59	504	481	23	248	238	10
1978	835	728	107	506	468	38	247	242	5
1979	1,045	862	183	529	475	54	236	228	8
1980	1,303	1,076	227	598	539	59	273	259	14
1981	1,434	1,164	270	666	587	79	234	217	17
1982	1,696	1,372	324	632	560	72	255	238	17
1983	1,392	1,104	288	672	567	105	268	238	30
1984	1,355	1,033	322	726	605	121	271	245	26
1985	1,276	990	286	713	600	113	303	271	32
1986	1,259	924	335	810	673	137	305	281	24
1987	1,152	854	298	765	600	165	392	347	45
1988	1,211	891	320	749	597	152	374	341	33
1989	1,114	853	261	815	634	181	380	335	45
1990	1,166	895	271	802	650	152	440	391	49
1991	1,166	912	254	787	607	180	489	412	77
1992	1,091	846	245	796	653	143	485	416	61
1993	1,216	956	260	849	682	167	535	449	78
1994	1,106	866	240	910	723	187	539	452	83
1995	1,046	799	247	852	668	184	588	489	95
1996	1,004	781	223	774	599	175	574	483	84
1997	1,063	804	259	724	550	174	582	470	106
1998	1,007	772	235	698	528	170	565	477	84
1999	NA	NA	NA	NA	NA	NA	469	376	88
2000	972	704	268	759	558	201	451	367	83
2001	930	667	263	709	536	173	497	392	105
2002	933	648	285	630	459	171	396	315	80
2003	950	674	276	720	537	183	474	373	101
2004	865	595	270	800	601	199	509	419	90

NOTE: NA, not available. Detailed national data were not released by the National Center for Education Statistics for the academic year ending in 1999. Some numbers may not sum to total because information on the gender of some recipients was missing.
SOURCE: Data from the National Science Foundation's (NSF's) *Science and Engineering Indicators*. Available at http://www.nsf.gov/statistics/nsf07307/pdf/tab53.pdf.

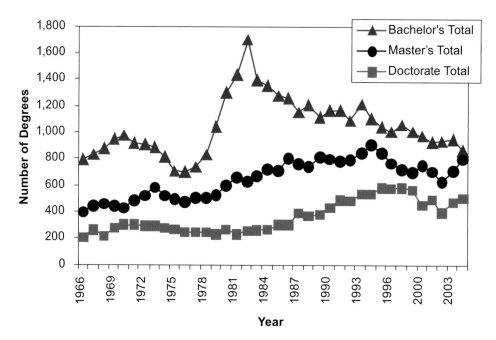

FIGURE 3-5 Materials and metallurgical engineering degrees awarded in the United States, by degree level. Data from NSF *Science and Engineering Indicators.*

CONCLUSION 7. Industry compensates for the inadequate corrosion engineering education of practicing engineers through on-the-job training and short courses. These skills- and knowledge-based, continuing-education approaches are widely accepted as useful and they play an important role, depending on the job function and desired outcomes. However, continuing education of the workforce is not a substitute for including corrosion in the curricula for graduating engineers and technologists.

CONCLUSION 8. The current university-based and on-the-job training approach to corrosion education does not allow the country to continue to reduce substantially the national cost of corrosion or to improve the safety and reliability of the national infrastructure. The corrosion engineering education system needs to be revitalized through (1) a series of shorter-term tactical actions by educators, government, industry, and the broader technical community and (2) longer-term strategic actions by the federal government and the corrosion research community.

RECOMMENDATIONS FOR A PATH FORWARD

Action is needed to improve the corrosion education of graduating and practicing engineers in the United States. Corrosion needs to be included in bachelor's-level design courses that are taught in the major engineering disciplines. Short courses and in-house training by corrosion experts have a role to play in increasing the number of corrosion-knowledgeable engineers, and savings will accrue when they apply their newfound knowledge, but these approaches have their limitations and are not an answer to inadequate education at universities that award professional degrees.

It is clear to the committee that students in general are not attracted to corrosion—not only because of the often negative connotations associated with the discipline but also because the jobs associated with corrosion are typically thought of as being jobs in basic industry. Corrosion is not thought to be relevant to or competitive with attractive fields such as nanotechnology, biomedical engineering, or jobs that tackle the energy crisis.

While the committee can envision that new approaches to distance learning might improve the state of corrosion education, it acknowledges that institutional barriers to distance learning will make it difficult in the near term. Also, linking corrosion to other electrochemical technologies or courses could enhance the status of corrosion and help to get it incorporated into existing curricula. Corrosion courses could be taught by experts in related areas such as electrochemistry.

The committee's tactical recommendations have several themes. One is that the corrosion education system depends on a substantial corps of corrosion teachers, which in turn depends on the health of the corrosion research community. A vibrant corrosion research community provides teachers for corrosion classes and a research environment in which corrosion students at all levels can gain experience. The corrosion teachers themselves will need to rely on the development of educational modules, case studies, and capstone courses, as well as on support for their own training and education.[9] Opportunities to gain work experience and to add to their knowledge in the government and industrial sectors is an important common theme. Another is the need for government and industry to identify and make known, on an ongoing basis, their corrosion skills requirements, thereby enabling the identification and updating of learning outcomes—that is, the skills a student is expected to have at graduation—for new educational programs (illustrative learning outcomes for some corrosion engineering courses are shown in Appendix F).

---

[9]A capstone course is a course offered in the final semester of a student's major. It ties together the key topics that faculty expect the student to have learned during the major, interdisciplinary program, or interdepartmental major.

Strategic Recommendations

During the course of the study the committee became convinced that two compelling challenges remained outstanding: one for the federal government, in particular DOD, and one for the corrosion community itself.

The first strategic recommendation is addressed to DOD, and specifically to its Corrosion Policy and Oversight Office. The committee is convinced that improving the education of the corrosion workforce, broadly defined, will hinge on the government's development of a strategic plan for fostering corrosion education with a well-defined vision and mission. An essential element of the plan will be how government can provide incentives to the educational sector to expand and revitalize corrosion engineering education. This plan will require input from a broad set of stakeholders and analysis and support from the government, industry, and academia.

Strategic Recommendation to the Government

The DOD's Director of Corrosion Policy and Oversight, whose congressionally mandated role is to interact directly with the corrosion prevention industry, trade associations, other government corrosion-prevention agencies, academic research and educational institutions, and scientific organizations engaged in corrosion prevention, should (1) set up a corrosion education and research council composed of government agencies, industry, and academia to develop a continuing strategic plan for fostering corrosion education and (2) identify resources for executing the plan. The plan should have the following vision and mission:

- *Vision.* **A knowledge of the environmental degradation of all materials is integrated into the education of engineers.**
- *Mission.* **To provide guidance and resources that will enable educational establishments to achieve the vision.**

The challenge to the corrosion community is based on the committee's observation that the community appears isolated from the larger scientific and engineering community. Repeatedly the committee heard that the general research and engineering community considers that corrosion science and engineering is a mature field, implying that there is little compelling science to be done. The committee heard from those who have studied corrosion for many years how many compelling science questions remain unanswered and how great the promise is for advancing corrosion mitigation and prevention if those questions can be answered.

Even small changes in environment and materials can adversely affect corrosion resistance and result in catastrophic degradation.

The committee is convinced that the responsibility for rectifying this faulty perception falls to the corrosion community itself. The education of a corrosion-savvy workforce is, broadly speaking, dependent on the health of the corrosion community, so the committee's second recommendation is addressed to that community.

Strategic Recommendation to the Corrosion Community

To build an understanding of the continuing need for corrosion engineering education, the corrosion research community should engage the larger science and engineering community and communicate the challenges and accomplishments of the field. To achieve this goal the corrosion research community should identify and publish the opportunities and priorities in corrosion research and link them to engineering grand challenges faced by the nation. To show how the field of corrosion could meet these challenges, the corrosion research community should reach out to its peers by speaking at conferences outside the field, publishing in a broad range of journals, and writing review articles for broad dissemination.

Tactical Recommendations

The committee presents its tactical recommendations in three ways: (1) by stakeholder—first, government, industry, and professional societies and, second, the university and education sector, (2) in the form of a summary matrix, and (3) by educational goal—namely, strategies for improving the education of identified segments of the engineering workforce.

By Stakeholder

To the Government, Industry, and Professional Societies

- Industry and government agencies, such as DOD, the Army Corps of Engineers, the Federal Highway Administration, state departments of transportation, DOT, and the U.S. Bureau of Reclamation, should strengthen the provision of corrosion courses and support the promulgation of corrosion-related learning outcomes by disseminating skills sets for corrosion technologists and engineers. The skills sets should be tied to actual case histories. Such an ongoing effort would enable the setting and periodic updating of learning outcomes for corrosion courses.

- Industry and federal government agencies, such as DOD's Office of Corrosion Policy and Oversight, the NSF, and the Department of Energy (DOE), should help develop a foundational corps of corrosion faculty by supporting research and development in the field of corrosion science and engineering. Such support could include the establishment of centers of expertise at key universities or in consortia of universities.
- Industry and federal government agencies, such as DOD's Office of Corrosion Policy and Oversight, should give universities incentives, such as endowed chairs in corrosion control, to promote their hiring of corrosion experts. The new DOD Faculty Fellowship follows this model.
- The DOD Office of Corrosion Policy and Oversight and NSF should support faculty development by facilitating their participation in research internships, short courses, and conferences.
- Industry and government agencies should partner with MSE and engineering departments to offer corrosion-related internships and sabbatical opportunities for students and faculty, respectively.
- Industry and federal government agencies, such as DOD, NSF, and DOE, should support graduate fellowships in corrosion engineering. As part of this effort, the DOD Office of Corrosion Policy and Oversight should establish a research support program equivalent to NSF educational experience programs, whereby a block grant awarded to an MSE or engineering department would fund some graduate students in the corrosion subspecialty.
- Federal government agencies, such as DOD's Office of Corrosion Policy and Oversight and DOE, should fund cooperative programs between university engineering and MSE departments and government laboratories to facilitate the graduate student research experience.
- Professional societies, such as NACE International and TMS, and government-supported materials research centers, such as NSF's Materials Research Science and Engineering Research Centers, should develop and provide materials for MSE and engineering departments that do not offer courses on corrosion engineering or do not have instructors with the relevant expertise. These educational modules would help nonexperts to deliver effective corrosion education. Such modules should be geared toward different areas of engineering—for example, biomedical, chemical, civil, mechanical, nuclear, and electrical engineering—and should include Web-based classes, problems, and case studies.
- Federal government agencies, such as DOD's Office of Corrosion Policy and Oversight and NSF, should fund the development of educational modules, including case studies and capstone courses, for use at community colleges and by university MSE and other engineering departments.
- Industry and government agencies should increase support for their engi-

neers to participate in short courses when specific skills shortages are identified and need to be remedied in a short time. These efforts will improve the knowledge and awareness of corrosion control on the part of practicing engineers and minimize their need for on-the-job training.

- The National Council of Examiners for Engineering and Surveying, with appropriate input from the professional societies, should tighten the requirement for corrosion in exams to certify professional engineers.

To the University and Education Sector

- Engineering departments in universities should incorporate electives and course work on corrosion into all engineering curricula. Improving the overall awareness of corrosion control will require that more engineers have basic exposure to corrosion, at least enough to "know what they don't know."
- MSE departments in the universities should set required learning outcomes for corrosion into their curricula. All MSE undergraduate students should be required to take a course in corrosion control so as to improve the corrosion knowledge of graduating materials engineers.
- Community colleges should add learning outcomes courses on corrosion engineering at the associate's level to provide technologists with a more specialized (industry- or application-specific) knowledge of corrosion.
- MSE and engineering departments at universities should provide continuing education in corrosion for practicing engineers.
- MSE and engineering departments in universities should provide the faculty to teach corrosion. To identify faculty with the appropriate expertise when no corrosion experts are on staff, departments should consider faculty who are expert in areas such as batteries and fuel cells, surface science, condensed matter physics, nanotechnology, and electrodeposition. The departments should also support participation in faculty development programs aimed at increasing the teaching capacity in corrosion.
- MSE departments at universities offering a required course in corrosion should ensure that they can continue to teach corrosion by hiring new faculty to replace retiring faculty who are experts in corrosion.
- MSE and engineering departments should partner with industry to create industry-guided capstone design courses for undergraduate engineers.

In Matrix Format

The tactical recommendations have just been listed by stakeholder or actor. Table 3-3 summarizes them in another way, as a matrix of recommended actions.

TABLE 3-3 Matrix of Recommended Actions

	Faculty Development	Curricula and Pedagogy
Industry	Should provide incentives to the universities, such as endowed chairs in corrosion control, to promote their hiring of corrosion experts. Should partner with universities to offer corrosion-related sabbatical opportunities for faculty.	Should partner with universities to create industry-guided capstone design for corrosion courses for undergraduate engineering students. Should strengthen the provision of corrosion courses by disseminating skills sets for corrosion technologists and engineers. Should partner with universities to offer corrosion-related internships for students.
Federal government	Should provide incentives to the universities, such as endowed chairs in corrosion control, to promote their hiring of corrosion experts. Should support faculty development, including participation in research internships, short courses, and conferences.	Should strengthen the provision of corrosion courses by publishing and publicizing skills sets for corrosion technologists and engineers. Government-supported research centers, such as those funded by DOE and NSF, should develop and provide materials for MSE and engineering departments that do not offer courses on corrosion engineering or do not have instructors with relevant expertise.

Teaching and Student Support	Research and Development	Workforce Development
Should support graduate student fellowships in corrosion engineering.	Should help develop a foundational corps of corrosion faculty by supporting research and development in the field of corrosion science and engineering.	Should increase support for the participation of their engineers in short courses.
Should support graduate student fellowships in corrosion engineering by establishing block grants to fund a number of graduate students in the corrosion subspecialty. Should fund cooperative programs between universities and government laboratories to facilitate the graduate student research experience. Should fund the development of educational modules—including case studies and capstone courses—for use by faculty at community colleges and university.	Should help develop a foundational corps of corrosion faculty by supporting research and development in the field of corrosion science and engineering.	Should increase support for the participation of their engineers in short courses.

continues

TABLE 3-3 Continued

	Faculty Development	Curricula and Pedagogy
University MSE departments	Should support participation in faculty development programs aimed at increasing the teaching capacity in corrosion. Should ensure adequate faculty and educational facilities are available to teach future corrosion experts by hiring new faculty and replacing retiring faculty who are experts in corrosion.	Should adopt required learning outcomes for corrosion in undergraduate curricula. Should require all MSE undergraduate students to take a course in corrosion control. Should partner with industry to create industry-guided capstone design for corrosion courses for undergraduate engineering students.
University engineering departments		Should adopt elective learning outcomes for corrosion in undergraduate curricula. Engineering departments should incorporate a corrosion course into all engineering curricula as an elective. Should partner with industry to create industry-guided capstone design for corrosion courses for undergraduate engineering students.
Professional societies		Professional societies should develop and provide materials for MSE and engineering departments that currently do not include courses on corrosion engineering or do not have instructors with relevant expertise.
Community college		Should adopt learning outcomes on corrosion in curricula for associates. Should add courses on corrosion engineering at the associates-degree level to provide technologists with better specialized (industry- or application-specific) corrosion knowledge.

Teaching and Student Support	Research and Development	Workforce Development
		Should provide corrosion continuing education courses for practicing engineers.
Should provide the faculty to teach corrosion. To identify faculty with the expertise to provide corrosion instruction when no corrosion experts are on staff, departments should consider faculty who are expert in areas such as batteries and fuel cells, surface science, condensed matter physics, nanotechnology, and electrodeposition.		Should provide corrosion continuing education courses for practicing engineers.
		The National Council of Examiners for Engineering and Surveying, with appropriate input from the professional societies, should tighten the requirement for corrosion in the relevant exams to certify professional engineers.

By Educational Goal

The committee also has broken down its tactical recommendations and looked at them yet another way—namely, as strategies for improving the education of (1) technologists, (2) non-MSE bachelor's-level engineering graduates, (3) MSE bachelor's-level graduates, (4) practicing engineers with bachelor's degrees, and (5) master's-level or Ph.D. students. Each strategy identifies actors, actions, and goals as appropriate.

Technologists

To provide technologists with better specialized (industry- or application-specific) knowledge,

- Community colleges should add courses on corrosion engineering at the associates degree level.
- Industry and government agencies, such as DOD, the Army Corps of Engineers, and the Bureau of Reclamation, should help to increase the availability of such courses by disseminating the skills sets needed by corrosion technologists. The skills sets should be tied to actual case histories. Such an ongoing effort would enable the setting and periodic updating of learning outcomes for such technologists.
- Industry should support corrosion technology programs at community colleges by providing internship opportunities.
- The federal government should fund the development of corrosion control educational modules for use by faculty at community colleges.
- Professional societies should provide corrosion technical courses and certification support.

Non-MSE, Bachelor's-Level Engineering Graduates

To improve the overall awareness of corrosion control among all graduating engineers, so that all engineers have a basic exposure to corrosion, enough to "know what they don't know,"

- Engineering departments in universities should incorporate a corrosion course into all engineering curricula as an elective.
- Industry and government agencies, such as DOD, the Army Corps of Engineers, and the Bureau of Reclamation, should help to increase the availability of such courses by disseminating skills sets for non-MSE engineers. The skills sets should be tied to actual case histories. Such an ongoing effort

would enable the setting and periodic updating of learning outcomes for corrosion-aware engineers.

- The National Council of Examiners for Engineering and Surveying, with appropriate input from the professional societies, should tighten the requirement for corrosion in exams to certify professional engineers.
- Industry and government should partner with university programs to offer corrosion-related internships and sabbatical opportunities for students and faculty, respectively.
- DOD and the NSF should provide financial support to university faculty who wish to attend short or summer courses to improve their ability to teach corrosion.
- Universities should offer and support their staff's participation in faculty development programs aimed at increasing the capacity to teach corrosion in their engineering departments.
- Professional societies, such as NACE International and TMS, and government-supported materials research centers, such as NSF's Materials Research Science and Engineering Centers (MRSECs), should provide supplementary course material for engineering curricula that currently do not cover corrosion and for engineering departments that do not have instructors with relevant expertise, by developing educational modules to assist nonexperts in delivering effective corrosion education. Such modules should be geared to different areas of engineering—for example, biomedical, chemical, civil, mechanical, nuclear, and electrical engineering—and should include Web-based classes, problems, and case studies.
- DOD and NSF should support the strengthening of corrosion engineering education in engineering departments by funding the development of educational modules, case studies, and capstone courses.
- Engineering departments in universities should also supply the faculty to teach corrosion. To identify faculty with the expertise to do that, programs should consider faculty who are expert in areas such as batteries and fuel cells, surface science, condensed matter physics, nanotechnology, and electrodeposition.

MSE Bachelor's-Level Graduates

To improve the corrosion knowledge of graduating materials engineers,

- MSE departments in the universities should require all MSE students to take a course in corrosion control.
- Industry and government agencies, such as DOD, the Army Corps of Engineers, and the Bureau of Reclamation, should help to increase the avail-

ability of such courses by publishing skills sets for MSE engineers. The skills sets should be tied to actual case histories. Such an ongoing effort would enable the setting and periodic updating of learning outcomes for corrosion-knowledgeable materials engineers.

- Industry should partner with MSE departments to create industry-guided capstone design courses.
- DOD and the NSF should support the strengthening of education in corrosion for materials engineers by funding faculty development, the development and provision of teaching materials, and supporting fellowships. Faculty development should include participation in research internships, short courses, and conferences.
- Professional societies, such as NACE International and TMS, and government-supported materials research centers, such as NSF's MRSECs, should develop and provide materials for MSE curricula that currently do not cover corrosion engineering and for MSE departments that do not have instructors with relevant expertise. These educational modules would assist nonexperts in delivering effective corrosion education to MSE students.
- MSE departments in universities should also provide the faculty to teach corrosion. To identify faculty with the expertise to provide corrosion instruction, programs should look for faculty who are expert in areas such as batteries and fuel cells, surface science, condensed matter physics, nanotechnology, and electrodeposition.
- Industry and government should partner with university programs through corrosion-related internships and sabbatical opportunities for students and faculty, respectively.

Practicing Engineers with Bachelor's Degrees

To improve the knowledge and awareness of corrosion control among practicing engineers and to minimize their need for on-the-job training,

- MSE departments at universities and technical professional societies, such as NACE and TMS, should provide corrosion courses for working professionals.
- Industry and government agencies, such as DOD, the Army Corps of Engineers, the Bureau of Reclamation, and others, should help to increase the availability of such courses by publishing descriptions of corrosion-related skills needed by the engineers in their workforce. The skills sets should be tied to actual case histories. Such an ongoing effort would enable the setting and periodic updating of learning outcomes for targeted short courses.
- Industry and government should support the participation of their engi-

neers in short courses when specific skills shortages are identified and must be filled in the short term.

Graduate Engineering Students

To increase the availability of corrosion expertise,

- MSE and engineering departments at universities should ensure that adequate faculty and educational facilities are available to teach future corrosion experts by hiring new faculty to replace retiring faculty who are experts in corrosion.
- Industry and federal government agencies, such as DOD, NSF, and DOE, should help develop a foundational corps of corrosion faculty by supporting research and development in corrosion science and engineering. Such support should include graduate fellowships and could include the development of Centers of Expertise (COEs) at key universities or in consortia of universities.
- Federal government agencies, such as DOD and DOE, should fund cooperative programs between universities and government laboratories to facilitate graduate student experience.
- Industry and the federal government agencies, such as DOD's Office of Corrosion Control, should provide incentives to the universities, such as endowed chairs, to promote their hiring of corrosion experts. The new DOD Faculty Fellowship follows this model.
- The DOD Office of Corrosion Control should establish a research support program equivalent to an NSF educational experience, whereby a block grant is awarded to fund a number of graduate students in the corrosion subspecialty at a university.

Appendixes

Appendix A

Two Earlier Reports

The first report summarized in this appendix is the 2001 study *Corrosion Costs and Preventive Strategies in the United States.*[1] The U.S. Federal Highway Administration (FHWA) commissioned the study because it wished to have quantified the economic impact of metallic corrosion, and it asked the authors to recommend preventive strategies to minimize the impact of corrosion on the U.S. economy. The second report summarized here is *Corrosion Control,* from the Defense Science Board (DSB), which released it in 2004.[2]

COST OF CORROSION STUDY

*Corrosion Costs and Preventive Strategies in the United State*s was written by CC Technologies Laboratories, Inc., from 1999 to 2001 under a cooperative agreement with the FHWA and NACE International. The study showed that although technological advances have brought about new ways of corrosion mitigation and better corrosion management techniques, the United States has a long way to go in optimizing corrosion control practices, and it must move toward better corrosion management using preventive strategies in both nontechnical and technical areas.

---

[1] For a copy of the report see http://www.corrosioncost.com/pdf/main.pdf. Accessed October 2008. For further information on the study see http://www.corrosioncost.com/home.html. Accessed October 2008.

[2] For a copy of the report see http://www.acq.osd.mil/dsb/reports/2004-10-Corrosion_Control.pdf. Acccessed October 2008.

Two obstacles exist to the development of advanced technologies for corrosion control and the implementation of those technologies: one is the general lack of awareness on the part of the public and policy makers of corrosion costs and the potential savings and the other is the widespread misconception that nothing can be done about corrosion.

The study points out that change is required in three areas: (1) the policy and management framework for effective corrosion control, (2) the science and technology of corrosion control, and (3) the transfer of technology and implementation of effective corrosion control. The authors believed that a national agenda should be carried out to reduce the economic impact of corrosion. One of the outstanding challenges to achieving this involves spreading the awareness of corrosion and the expertise needed to deal with it, both of which are currently scattered throughout government and industry organizations.

Approaches and Results

The study employed two approaches to estimate the total cost of corrosion. Approach 1 (corrosion control methods and services) involved estimating the total cost of corrosion to the economy by summing up the cost of corrosion products and services. The corrosion control methods included protective coatings, corrosion-resistant alloys, corrosion inhibitors, polymers, anodic and cathodic protection, and corrosion control and monitoring equipment. Corrosion control services included engineering research and development and education and training. The total cost of corrosion control methods and services was estimated at $121 billion, or 1.38 percent of the $8.79 trillion U.S. gross domestic product (GDP) in 1998. The cost of organic coatings, $107.2 billion, constituted approximately 88 percent of the total cost of corrosion control methods and services. The cost of corrosion control services contributed less than 1.2 percent to the total cost. Past studies have shown that this approach, though admittedly simple, is likely to miss the significant cost of corrosion management, the cost of direct services related to the owner/operator, and capital losses due to corrosion.

Approach 2 (industry sector analysis) consisted of first estimating the costs of corrosion in 26 specific industry sectors, each of which had been assigned to one of five broader categories, and then extrapolating them to obtain a nationwide estimate of total corrosion cost. The sectors were selected to represent as broad a cross section of the U.S. economy as possible. All had their specific corrosion problems, and together they constituted approximately 27 percent of the U.S. GDP. The five major sector categories were infrastructure, utilities, transportation, production and manufacturing, and government. Data collection methods for sector-specific analyses differed significantly from sector to sector, with the sources ranging from government reports, publicly available documents, expert opinion, and industry records. Below are the five main categories and the sectors that were assigned to them:

- *Infrastructure.* Highway bridges, gas and liquid transmission pipelines, waterways and ports, hazardous materials storage, airports, and railroads.
- *Utilities.* Gas distribution, drinking water and sewer systems, electrical utilities, and telecommunications.
- *Transportation.* Motor vehicles, ships, aircraft, railroad cars, and hazardous materials transportation.
- *Production and manufacturing.* Oil and gas exploration and production; mining; petroleum refining; chemicals; petrochemicals and pharmaceuticals; pulp and paper; agriculture; food processing; electronics; and home appliances.
- *Government.* Defense and nuclear waste storage.

The costs in each of the five categories were discussed next:

- *Infrastructure.* The annual direct cost of corrosion in the infrastructure category was estimated at $22.6 billion and accounted for approximately 16.4 percent of the total cost of the five categories. The most significant contributions in this category were from the industrial sectors of highway bridges ($8.3 billion), gas and liquid transmission pipelines ($7 billion), and hazardous materials storage ($7 billion). Corrosion costs for the airports and railroad sectors were not estimated owing to insufficient information.
- *Utilities.* The annual cost of corrosion in the utilities category was estimated to be $47.9 billion and represented 34.7 percent of the total cost of the five categories. The industrial sector of drinking water and sewer systems was the single largest contributor, with an annual corrosion cost of $36 billion, followed by electrical utilities and gas distribution, which had annual corrosion costs of $6.9 billion and $5 billion dollars, respectively. No estimate of corrosion cost was made for the telecommunications sector because there was little information on this rapidly changing industry.
- *Transportation.* The annual cost of corrosion cost in this category was estimated at $29.7 billion, which accounted for 21.5 percent of the total cost of the five categories. The transportation category comprised the industrial sectors of motor vehicles, ships, aircraft, railroad cars, and hazardous material transport. The motor vehicles sector was estimated to have the largest corrosion costs, $23.4 billion.
- *Production and manufacturing.* This category was made up of the sectors that produce goods of crucial importance to the economy. The annual cost of corrosion in this category was estimated at $17.6 billion, which was

12.8 percent of the total cost of five categories. The pulp and paper sector had the largest corrosion cost, $6.0 billion, in this category, followed by the petroleum refining sector, with a corrosion cost of $3.7 billion. No estimate of corrosion cost was made for the electronics sector because it is difficult to detect and identify corrosion failures.

*   *Government.* Defense and nuclear waste storage are the two sectors analyzed in this category. The total annual cost of corrosion in the category was estimated to be $20.1 billion, representing approximately 14.6 percent of the total cost of the five categories analyzed. The defense sector, with an annual corrosion cost of $20.0 billion, accounted for 99.5 percent of the total.

Since the total cost of $137.9 billion represented only the estimated direct corrosion cost for the analyzed industrial sectors, the total direct cost of corrosion to the U.S. economy was estimated using a nonlinear extrapolation to obtain an estimate for the nation of $276 billion, which was 3.1 percent of the $8.79 trillion U.S. GDP for the year 1998. The indirect corrosion costs, i.e., the costs incurred by those other than the owners and operators, were conservatively estimated to be the same as the direct costs, resulting in a direct plus indirect cost of $512 billion, which was approximately 6.2 percent of the U.S. GDP for 1998. The large indirect costs included (1) lost productivity because of outages, delays, failures, and litigation; (2) taxes and overhead on the cost of corrosion portion of goods and services; and (3) indirect costs of nonowner/operator activities.

The difference in the costs estimated by Approach 1 ($121 billion) and Approach 2 ($512 billion) can be attributed to the fact that the former included only corrosion control materials and services. The latter, however, also took into account owner/operator corrosion management costs and indirect costs that were not part of the corrosion services cost.

Conclusions and Recommendations of the 2001 Report

The nation's infrastructure is essential to the quality of life, industrial productivity, international competitiveness, and security. Each component of the infrastructure, such as highways, airports, water supply, waste treatment, energy supply, and power generation, is a complex system and a significant investment. Corrosion is a primary cause of damage and poses a great threat to the infrastructure. The study points out that technological advances in the last several decades have given us ways to prevent and manage corrosion. A number of cost-effective corrosion management techniques can significantly enhance the service life of existing systems and avoid the need for new construction and replacements. Preventive strategies have been identified in both nontechnical (management and public policy) and technical areas. In nontechnical areas, they include the following:

- Increase awareness of the significant costs of corrosion and the potential for savings when it can be prevented.
- Correct the misconception that nothing can be done about corrosion.
- Change policies, regulations, standards, and management practices to increase cost savings through sound corrosion management.
- Improve the education and training of staff in the recognition and control of corrosion.

Preventive strategies in technical areas include advances in design practices for better corrosion management; in life prediction and performance assessment methods; and in corrosion technology generally through research, development, and implementation.

It will be necessary to engage more of the primary stakeholder—government and industry, the general public, and consumers—in this effort and to harness the awareness and expertise that is currently scattered throughout government and industry organizations. The following recommendations were made:

- Form a committee on corrosion control and prevention under the National Research Council. The focus of the committee would be to preserve and extend the life of existing infrastructure and equipment.
- Develop a national focus on corrosion control and prevention.
- Improve policies and corrosion management.
- Work for technological advances that would allow reducing the cost of corrosion.
- Implement effective corrosion control.

DSB REPORT ON CORROSION CONTROL

The readiness and safety of weapons systems are among the highest priority challenges for DOD. It is imperative that the materials of which DOD equipment is made be maintained in an acceptable condition so that the equipment can be employed safely as soon as it is required, including in harsh and physically demanding environments. The dollar cost of corrosion to DOD has been estimated by the Government Accountability Office (GAO) to be $10 billion to $20 billion per year. At the request of the Acting Under Secretary of Defense for Acquisition, Technology and Logistics and the Deputy Under Secretary of Defense for Logistics and Materiel Readiness, the DSB formed a task force to address corrosion control efforts in DOD.

The task force made five recommendations, one in each area that represents a barrier to improvement: (1) leadership commitment and policy, (2) design and manufacturing practices, (3) maintenance practices, (4) funding and management, and (5) the scientific basis for preventing and mitigating corrosion.

Scope and Approach

The task force was specifically asked to do these things:

- Assess ongoing corrosion control efforts within DOD, with specific attention to:
  —Duplication of research efforts.
  —Application of existing and future technology that exists in one area to other areas.
  —Status of training for operator and maintenance personnel.
  —Status of maintenance processes.
  —Incorporation of corrosion control and maintainability in ongoing acquisition programs.
  —Identification of unique environments important to national security but with few commercial applications.
- Determine which areas would provide the most significant advances in combat readiness if adequate resources are applied.
- Assess best commercial practices and their applicability.

Both DOD's infrastructure (facilities, bases, ports) and its weapons systems (platforms, electronics, munitions) experience corrosion, but only the weapons are directly involved in operational readiness and combat capability. Accordingly, and in line with the terms of reference, the task force directed most of its attention to corrosion problems associated with the weapons systems and equipment. In the absence of accurate corrosion cost data, it is impossible to quantify the potential benefits from serious corrosion management. Nonetheless, the consensus within the task force and among many DOD and industry experts holds that as much as 30 percent of the corrosion costs can be avoided by preventing rather than repairing. This is not a near-term target and is heavily dependent on reforms in the design of DOD weapons systems and in the acquisition process. It is likely that a great deal of corrosion prevention and treatment could be funded with the dollars now being used to rework or replace the most badly damaged items. The task force suggested that the cost and readiness impact of corrosion can be reduced if the DOD manages the corrosion challenge better. This, however, requires the removal of certain institutional barriers.

Results and Recommendations of the 2004 Report

The task force's examination of DOD's corrosion program revealed several opportunities for both short-term (tactical) and long-term (strategic) improvement. Many of them remain unrealized because of the barriers encountered.

Leadership Commitment and Policy

- The absence of any priority for serious attention to corrosion reflects the leadership's ignorance of the problem, an ignorance due not to incompetence but rather to a lack of accurate and meaningful data.
- DOD does not have accurate information on the direct and indirect costs of corrosion prevention, mitigation, and remediation. Consequently, it has no strategy for systemic improvement.
- Corrosion costs being unclear, service decision makers lack compelling arguments for resources to reduce life-cycle costs. The problem is that the system incentivizes minimum acquisition cost rather than life-cycle cost.
- Decision makers lack effective corrosion standards and test methods to assess corrosion performance.

Design and Manufacturing Practices

- Since design-phase decisions largely determine future corrosion costs, materials choice, coating selection, and structural aspects are critical. Corrosion specialists must participate, and advanced technologies must be considered.
- The predictive corrosion models needed for the design of weapons systems do not exist.
- There are no adequate maintenance cost accounting systems for estimating return on investment.
- Acquisition and design personnel have not been given the training they need to consider minimizing the impact of corrosion on life-cycle costs.
- Existing DOD standards and metrics are often advisory rather than mandatory.

Recommendation 1: Promulgate and enforce policy emphasizing life-cycle costs over acquisition costs in procurement and provide the incentives and training to assure that corrosion costs are fully considered in design, manufacturing, and maintenance. The total near-term investment cost on the order of $1 million is estimated for implementing this recommendation, primarily to assemble a standing team of corrosion experts to advise decision makers.

Recommendation 2: Mandate and implement comprehensive and accurate corrosion data reporting systems across DOD, using standard metrics and definitions. The total cost for implementing this recommendation is estimated at about $5 million, largely for contract support in the development of standards and metrics.

Maintenance Practices

- Maintenance needs and current state of corrosion are not well characterized for most nonaviation assets.
- It has been shown in industry that the costs of corrosion can be drastically reduced by instituting best practices in engineering and maintenance.
- Systematic corrosion control training and awareness are lacking on the part of operators and maintainers.
- There are no consistent, comprehensive corrosion control and maintenance strategies throughout the services and for all systems, including the infrastructure.

Recommendation 3: Fund contract for comprehensive assessment of all DOD weapon system equipment by approximately 30 five-person teams of corrosion experts and use the results to develop and implement a comprehensive corrosion maintenance strategy. The cost for implementing this recommendation is estimated at about $25 million per year and should be continued indefinitely.

Funding and Management

- Corrosion science and technology (S&T) funding is small, fragmented and generally comes from unrelated research and development accounts such as Small Business Innovative Research (SBIR) and the Strategic Environmental Research and Development Program (SERDP).
- Dollars devoted to corrosion prevention during weapon systems research, development, testing, and evaluation (RDT&E) have historically proved insufficient.
- No specific corrosion remediation budget exists in service operation and support (O&S) accounts.

Recommendation 4: Establish a Corrosion Executive for each service with responsibility for oversight and reporting, full authority over corrosion-specific funding, and a strong voice in corrosion-related funding. An investment of $60 million per service is estimated.

Scientific Basis for Prevention and Mitigation of Corrosion

- Long-term funding for corrosion S&T that would make the successful application of research more likely is too low by a factor of three or so.
- That there is little or no redundancy in corrosion S&T portfolios can be attributed to the diversity of issues and platforms across the various services.
- Current S&T portfolios are very technology-oriented and appear to be short on the kind of research that brings a better understanding of the science that underlies corrosion.
- Several S&T areas have no dedicated, consistent S&T program in corrosion.
- There is adequate communication across the corrosion S&T community, so duplication of efforts is not a problem.

Recommendation 5: Refocus and reinvigorate corrosion S&T portfolios; triple the effective funding in this area (+$20 million). It is estimated that an additional $20 million per year would be required.

The task force estimated the cost for implementing the above five recommendations at approximately $50 million in the first year, assumed to be FY2005. Once the foundations are laid in the first year, additional investment in preventive design in future years is estimated to be $100 million to $150 million per year but is expected to quickly (within 1 or 2 years) be offset by corresponding and larger reductions in O&S.

Appendix B

Data Gathered from Universities

Responses were received from 31 institutions[1]

Arizona State University
Arkansas State University
Brigham Young University
California Polytechnic State University
Gannon University
Iowa State University
Lafayette College
Lehigh University
Michigan State University
Mississippi State University
North Carolina State University
North Dakota State University
Northwestern University
The Ohio State University
Purdue University
San Jose State University
State University of New York (SUNY) Maritime College

---

[1]These data were gathered by means of an online questionnaire. While every attempt was made to get responses from a representative selection of institutions, respondents ultimately were self-selected. Accordingly, data should not be considered to constitute a comprehensive statistical survey.

Trine University (before 8/1/2008, Tri-State)
University of Alabama at Birmingham
University of California, Berkeley
University of Florida
University of Kansas
University of Maryland
University of Massachusetts, Lowell
University of Michigan
University of Pennsylvania
University of Puerto Rico
University of Texas at Austin
University of Texas at El Paso
University of Wisconsin-Madison
Virginia Polytechnic Institute and State University

Breakdown of schools/departments
- Materials science, 61%
- Other (chemical, civil, mechanical, environmental, or general engineering), 39%

Do you follow a quarter system or a semester system?
- Quarter, 10%
- Semester, 90%

Do you offer a course or courses specifically in corrosion?
- Yes, 61%[2]
- No, 39%

Level at which course is taught
- Graduate, 27%
- Undergraduate, 46%
- Mixed, 27%

Class size
- Fewer than 20, 77%
- 20-50, 19%
- More than 100, 4%

---

[2]See Table B-1 for a list of the 26 such courses.

Required or elective
- Required, 23%
- Elective, 77%

Frequency
- Yearly, 58%
- Every other year, 38%
- Less frequently than every other year, 4%

Majors requiring the course
- Metals specialty for a materials engineering degree
- Chemical engineering
- Metallurgy, biomaterials specialization in MSE
- Materials science/mechanical engineering joint degree

Do students from other departments enroll in the corrosion course?
- Yes, 61%
- No, 39%

What are these other departments (see Figure B-1)?
- Aerospace engineering
- Biological engineering
- Biomedical engineering (3)
- Chemical engineering (4)
- Chemistry
- Civil engineering (3)
- Civil and environmental engineering
- Dentistry
- Electrical engineering and computer science
- General engineering
- Mechanical engineering (5)
- Mining and materials engineering
- Nuclear engineering
- Petroleum engineering
- Welding

Do any of your courses address the electrochemical fundamentals of corrosion?
- Yes, 100%

Do any of your courses address ways to minimize corrosion by design?
- Yes, 100%

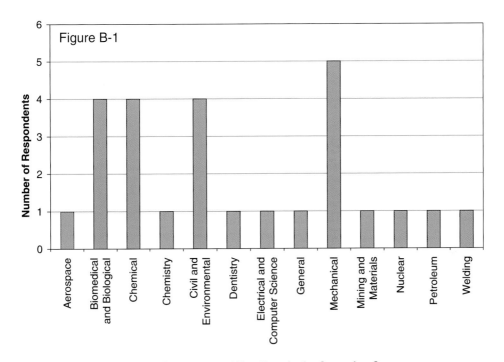

Home Department of Enrollees in the Corrosion Course

What are your reasons for offering a course in corrosion?

- Preparation of students for careers as practicing materials engineers must contain elements of the full life cycle of materials. This includes materials degradation.
- Student interest.
- Essential knowledge for a metallurgist.
- Critical for materials engineers.
- This is an important course for those working in the electronics industry.
- It is important for students to understand how to prevent corrosion. Also, the same concepts help them understand electroplating, batteries, and other phenomena.
- To equip our graduates with practical tools as an outcome.
- I have been associated with the topic for over 40 years and know of its impact and importance to technology and commerce.
- Crucial information for materials scientists and engineers and highly useful for other engineers

TABLE B-1 Corrosion Courses

Name of Course	Credits	Level at Which Course Is Taught	Required/ Elective	Department in Which Course Is Taught	Frequency	Number of Students
Corrosion	3	Graduate	Elective	Mechanical engineering	Yearly	<20
Corrosion	3	Mixed graduate/ undergraduate	Elective	Mechanical engineering	Yearly	<20
Corrosion	3	Mixed graduate/ undergraduate	Elective	Materials science	Every other year	<20
Corrosion	3	Mixed graduate/ undergraduate	Elective	Materials science	Every other year	<20
Corrosion	3	Undergraduate	Elective	Materials science	Yearly	<20
Corrosion	3	Undergraduate	Required	Materials science	Yearly	<20
Corrosion (Chemical Properties)	3	Undergraduate	Required	Materials science	Yearly	20-50
Corrosion and Corrosion Protection	1	Undergraduate	Elective	Materials science	Less frequently than every other year	<20
Corrosion and Failure Analysis	3	Undergraduate	Required	Materials science	Yearly	20-50
Corrosion and Oxidation	2	Graduate	Elective	Materials science	Every other year	<20
Corrosion Engineering	3	Graduate	Elective	Chemical and petroleum engineering	Yearly	<20
Corrosion Engineering	3	Undergraduate	Required	Materials science	Yearly	20-50
Corrosion Engineering	3	Undergraduate	Elective	Mechanical engineering	Yearly	<20
Corrosion Lab	1	Undergraduate	Elective	Materials science	Yearly	<20

TABLE B-1 Continued

Name of Course	Credits	Level at Which Course Is Taught	Required/ Elective	Department in Which Course Is Taught	Frequency	Number of Students
Corrosion Science	3	Graduate	Elective	Materials science	Yearly	<20
Degradation of Materials	3	Mixed graduate/ undergraduate	Elective	Materials science	Every other year	>100
Engineering Materials (Chemical Engineering)	3	Undergraduate	Required	Materials science	Yearly	20-50
Environmental Degradation of Materials	3	Mixed graduate/ undergraduate	Elective	Materials science	Yearly	<20
Environmental Effects on Materials Properties and Behavior	3	Graduate	Elective	Materials science	Yearly	<20
Industrial Corrosion	3	Mixed graduate/ undergraduate	Elective	Materials science	Every other year	<20
Kinetics of Materials Reactions	3	Undergraduate	Elective	Materials science	Every other year	<20
Principles of Corrosion	3	Mixed graduate/ undergraduate	Elective	Materials science	Yearly	<20
Principles of Corrosion and Electrochemical Processes	3	Graduate	Elective	Materials science	Every other year	<20
Principles of Materials Corrosion	3	Undergraduate	Elective	Materials science	Every other year	<20
Special Topics in Corrosion Science	3	Graduate	Elective	Materials science	Every other year	<20
Stability of Materials	3	Undergraduate	Required	Materials science	Yearly	20-50

- The two upper division courses that our graduates tell us are most important to their career are Corrosion and Failure Analysis. This feedback is in agreement with faculty perceptions. Both courses were switched from electives to required courses 3 years ago.
- Corrosion is an important aspect of materials engineering, and students often need this information to design components or solve failure problems.
- Undergraduate student interest in the subject.
- Student/faculty interest.
- Student demand, employer demand, distance education demand. Central part of the curriculum.
- Generally, corrosion is almost everywhere, and teaching students this topic broadens their knowledge of material selection, design against corrosion, and prevention of corrosion.
- Understanding of the fundamentals of corrosion. Techniques for testing for corrosion. Methods of preventing corrosion.
- It is an important topic for many applications.

Who teaches your corrosion courses? How are they trained?
- Tenured and tenure-track faculty. These faculty have Ph.D. degrees in materials science and industrial experience.
- Lecturer.
- Specialists in corrosion. They have large research activities in corrosion.
- Courses are taught by regular faculty whose specialty is physical metallurgy with an emphasis on corrosion/failure analysis.
- Ph.D.-level faculty.
- A full professor who does research on electrochemistry.
- Chemical engineering faculty members trained through formal course work and attending seminars related to industrial corrosion.
- Ladder-rank faculty whose doctorates and research experience is either corrosion or related electrochemical fields.
- Materials professor with background in electrochemistry/corrosion.
- A professor with no formal training in corrosion.
- I teach the course. I have a Ph.D. in metallurgy and materials science; over 40 years experience; taught the topic since 1965 at the college level.
- Strong background in applied chemistry and attended corrosion short course at MIT in early 1980s. Has taught the corrosion course for more than 20 years.
- Has a B.S. in materials engineering and a Ph.D. in biomedical engineering. Did biomaterials corrosion/wear research for his Ph.D.
- Ceramics researcher by training and metallurgy researcher by training.
- Electrochemical processes is a major interest area in his research.

Does the content of your corrosion course(s) cover a broad spectrum, or is it focused on satisfying the objectives required by the major/discipline?[3]

- Very broad spectrum, including metals, ceramics, polymers, wood, bio-materials, and biodegradable materials and unique environmental conditions, including marine, space, high temperature, high humidity, etc.
- The course is focused on corrosion in metals, with a focus on both making and designing structures based on existing materials and on the aging of materials already in service.
- Broad spectrum of materials and applications. We also cover materials degradation in other courses, such as polymer degradation in our polymer course.
- The course is focused on aqueous and atmospheric corrosion of metals.
- The content of this course includes forms of corrosion and their mechanisms and principles of electrochemistry for cathodic protection and electrodeposition.

How much has the content of your corrosion classes changed in the past 10 years?[4]

- It has changed significantly, primarily the inclusion of other degradation phenomena in addition to the classical electrochemical nature of aqueous metallic corrosion.
- Quite a bit. The course has migrated from being essentially all corrosion to include more failure analysis and less corrosion (more or less 50-50 split now).
- The instructor has changed, and there is now more of an emphasis on electrochemistry.
- Not very much, it is a sophomore-level undergraduate class so it focuses on the basics of corrosion.
- Fairly significantly, responding to new research findings and different topical applications.
- More emphasis on polymer degradation; microbiologically induced corrosion; impact of corrosion on fracture mechanics as it applies to stress corrosion and fatigue.
- The content has changed to include more electrochemical background and its application so that students gain an understanding of underlying principles rather than build knowledge through case studies.

---

[3]Most other schools simply answered "broad" without going into detail.
[4]Other schools answered "very little" or gave a number without details.

If your institution does not offer corrosion, why not?
- Other topics have more priority, 25%
- No one available for or interested in teaching it, 33%
- The material is covered in other courses, 42%

Is corrosion covered in other courses?[5]
- Yes, 80%
- No, 20%

Required or elective
- Required, 80%
- Elective, 20%

Class size
- Fewer than 20, 30%
- 20-50, 41%
- 51-100, 16%
- More than 100, 14%

Frequency of classes
- Yearly, 66%
- Every semester, 18%
- Every other year, 11%
- Every quarter, 5%

Time devoted to discussion of corrosion
- One lecture, 41%
- A few lectures, 48%
- Multiple lectures, 11%

Class level
- Undergraduate, 79%
- Mixed graduate/undergraduate, 19%
- Graduate, 2%

Majors that require the courses in which corrosion is discussed (see Figure B-2)
- Materials science and engineering (4)
- Bioengineering

---

[5]Table B-2 lists these other courses.

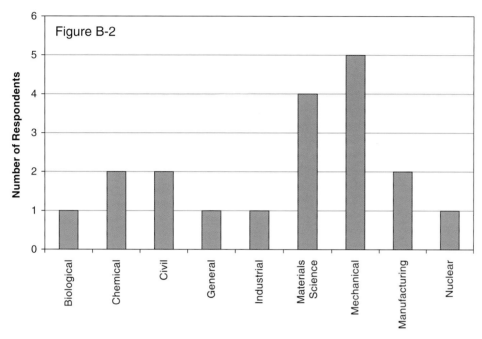

Majors Requiring One or More of the Courses That Teach Some Corrosion

- Mechanical engineering (5)
- Industrial engineering
- Chemical engineering (2)
- Civil engineering (2)
- Manufacturing engineering (2)
- General engineering
- Nuclear engineering

Which other departments teach the course?
- Chemistry (2)

Do students from other departments enroll in the course(s)?
- Yes, 57%
- No, 43%

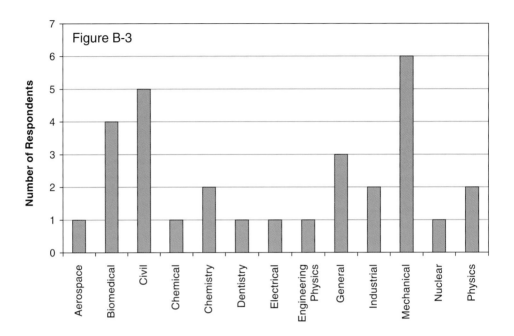

Home Department of Enrollees in Courses That Teach Some Corrosion

From which other engineering departments do students enroll in the course (Figure B-3)?

- Mechanical engineering (6)
- Chemical engineering (5)
- Chemistry
- General engineering (3)
- Nuclear engineering
- Biomedical engineering (4)
- Civil engineering (2)
- Industrial engineering (2)
- Dentistry
- Engineering physics
- Aerospace engineering
- Physics (2)
- Electrical engineering

If you do not offer corrosion in other courses why not?

- Other topics have more priority, 50%
- Nobody interested in or available for teaching it, 17%
- Other, 33%

Other reasons:
- Covered in the corrosion course
- There is an entire course devoted to it, so there is no need to introduce it in other courses.

Do you offer lab-based courses in corrosion?
- Yes, 11%[6]
- No, 89%

Would your department consider hiring a faculty member whose technical focus is corrosion?
- Yes, 59%
- No, 41%

Would this faculty member
- Fill a vacancy by a retiring or newly retired faculty member, 50%
- Fill a newly created slot with requisite facilities set aside for the new hire, 12.5%
- Other, 37.5%

Other answers:
- We would consider such a candidate if the candidate is competitive with candidates from other specialties. We do not have a specific position set aside for corrosion studies.
- If they have strong materials expertise, not just corrosion.
- If their work also involved applications of electrochemistry to energy production.
- Would consider, but are not seeking this expertise for a current PVL.
- Either vacancy by retirement or creating a new slot.

Why would you not consider hiring such a faculty member?
- Other topics have more priority, 91%
- Limited availability of research funds, 9%

Do you offer distance courses in corrosion?
- Yes, 7.5%[7]
- No, 92.5%

---

[6]Listed in Table B-3.
[7]Listed in Table B-4.

TABLE B-2 Other Courses in Which Corrosion Is Taught

Name of Course	Credits	Required/ Elective
Biomaterials	3	Elective
Chemistry for Mechanical Engineering	4	Required
Civil Engineering Materials	3	Required
Design I	3	Required
Design II	3	Required
Electrochemistry	3	Elective
Environmental Effects of Engineering Materials	3	Elective
Experimental Materials Science and Design	2	Required
Failure Analysis	3	Required
Ferrous Metallurgy	3	Required
Foundation Engineering	1	Required
Fundamentals of Materials Engineering	3	Required
Introduction to Ceramics	3	Required
Introduction to Materials	3	Required
Introduction to Materials Engineering	3	Required
Introduction to Materials Engineering Lab	1	Required
Introduction to Materials Science	3	Required
Introduction to Materials Science	3	Required
Introduction to Materials Science and Engineering	3	Required
Introduction to Materials Science and Engineering	5	Required
Introduction to Surface Science	3	Elective
Materials and Manufacturing for Aero and Ocean Engineers	3	Required
Materials Design	1	Required
Materials Science	3	Required
Mechanical Behavior of Materials	3	Required
Metallurgy	3	Required
Physical Ceramics	3	Required
Physical Materials II	3	Required
Physical Metallurgy	3	Required
Plant Design	3	Required
Polymer Technology and Engineering	3	Elective
Process Design	4	Required
Process Monitoring and Control	4	Required
Processing of Metallic Materials	3	Elective
Properties of Materials	3	Required
Steel Design	1	Required
Structural and Biomaterials	3	Elective
Structure and Properties of Materials	3	Required
Structure/Property Relations in Metals	3	Required
Thermodynamics	3	Elective
Thermodynamics	4	Required
Thermodynamics	3	Required
Thermodynamics in Materials Engineering	3	Required
Transport Phenomena	3	Elective

Graduate/ Undergraduate	Course is Offered	Enrollment	Time Devoted to Discussion of Corrosion
Mixed graduate/undergraduate	Every other year	<20	A few lectures
Undergraduate	Yearly	51-100	One lecture
Undergraduate	Yearly	<20	A few lectures
Undergraduate	Yearly	<20	A few lectures
Undergraduate	Yearly	<20	One lecture
Undergraduate	Yearly	<20	Multiple lectures
Undergraduate	Every other year	<20	Multiple lectures
Undergraduate	Yearly	20-50	One lecture
Undergraduate	Yearly	<20	A few lectures
Undergraduate	Yearly	20-50	A few lectures
Undergraduate	Every other year	<20	One lecture
Undergraduate	Yearly	20-50	A few lectures
Mixed graduate/undergraduate	Yearly	20-50	A few lectures
Undergraduate	Every semester	>100	One lecture
Undergraduate	Every quarter	>100	One lecture
Undergraduate	Every quarter	>100	One lecture
Undergraduate	Every semester	>100	One lecture
Undergraduate	Every semester	51-100	One lecture
Undergraduate	Yearly	51-100	A few lectures
Undergraduate	Yearly	20-50	A few lectures
Mixed graduate/undergraduate	Every other year	<20	One lecture
Undergraduate	Yearly	51-100	Multiple lectures
Undergraduate	Yearly	20-50	A few lectures
Undergraduate	Every semester	51-100	A few lectures
Undergraduate	Yearly	51-100	A few lectures
Undergraduate	Every semester	51-100	A few lectures
Undergraduate	Yearly	20-50	One lecture
Undergraduate	Every semester	20-50	Multiple lectures
Undergraduate	Yearly	20-50	One lecture
Mixed graduate/undergraduate	Yearly	20-50	A few lectures
Undergraduate	Yearly	20-50	A few lectures
Mixed graduate/undergraduate	Yearly	20-50	A few lectures
Undergraduate	Yearly	<20	A few lectures
Mixed graduate/undergraduate	Yearly	20-50	One lecture
Undergraduate	Every semester	>100	One lecture
Undergraduate	Yearly	20-50	One lecture
Mixed graduate/undergraduate	Yearly	<20	Multiple lectures
Undergraduate	Every semester	>100	One lecture
Undergraduate	Yearly	20-50	A few lectures
Graduate	Yearly	<20	A few lectures
Undergraduate	Yearly	20-50	A few lectures
Undergraduate	Yearly	20-50	A few lectures
Undergraduate	Yearly	20-50	One lecture
Mixed graduate/undergraduate	Every other year	<20	One lecture

TABLE B-3 Lab-Based Courses in Corrosion

Name	Credits	Graduate/ Undergraduate	Required/ Elective	Number of Sessions	Independent/Part of Another Course
Corrosion and Failure Analysis	3	Undergraduate	Required	16-20	Part of another course
Corrosion Engineering	3	Undergraduate	Required	21-25	Part of another course
Corrosion Lab	1	Mixed graduate/ undergraduate	Elective	6-10	Independent

TABLE B-4 Distance Courses in Corrosion

Name of Course	Credits	Graduate/ Undergraduate	Required/ Elective	Enrollment	Accept Nonenrolled Students?
Corrosion	3	Mixed graduate/ undergraduate	Elective	<20	Yes
Corrosion Engineering	3	Undergraduate	Elective	<20	Yes
Corrosion Science	3	Graduate	Elective	<20	Yes

How are nonenrolled students allowed to take the course?

- This option is usually available to students at companies with which we coordinate distance-education offerings. We consider these students as prospects for admission to formal degree programs. They must hold a relevant undergraduate degree and, in general, be admissible to the graduate school.
- With permission of the instructor.

Do you offer short professional courses in corrosion?

- Yes, 3%[8]
- No, 96%

Are there any prerequisites to taking these short professional courses?

- No, 100%

---

[8]Information on the single short course may be found in Table B-5.

TABLE B-5 Short Course on Corrosion

Name of Course	Credits	Graduate/ Undergraduate	How Often Offered	Enrollment	Where Taught
Special Topics in Corrosion	1	Open enrollment	Infrequently	<20	On campus

What percentage of your bachelor's-level graduates wind up working in the following fields (average across responses)?

- Design, 23%
- Manufacturing, 40%
- Research or academia, 24%
- Other, 13%

What percentage of your master's-level graduates wind up working in the following fields (average across responses)?

- Design, 21%
- Manufacturing, 34%
- Research or academia, 38%
- Other, 7%

What percentage of your Ph.D.-level graduates wind up working in the following fields (average across responses)

- Design, 14%
- Manufacturing, 23%
- Research or academia, 56%
- Other, 7%

Optional Questions

Is your department doing any corrosion-specific research? If so, who is funding it?

- Yes. DOE, DOD, SERDP, industry,
- Yes. Funded by Air Force.
- DOE.
- DOD/industry/DOE.
- No, although some corrosion-prevention work was done in the past.
- Industry, DOE, NSF, state.
- Federal Aviation Administration, Department of Defense.
- Not much now but anticipate substantially more soon. DOD.

Do you have any actual or potential partnerships with industry to study corrosion or develop continuing education for practicing engineers? If so, please describe them.

- Yes. Research activities.
- Yes. Joint projects, industry-sponsored projects, short courses onsite for engineers.
- Have a nearly completed plan to establish a distance education corrosion course in cooperation with ASM.
- Ongoing industrial partnerships in areas such as oil pipelines, semiconductor manufacturing, etc.
- We are developing continuing education options with a number of manufacturing companies.

How does the teaching of corrosion R&D fit into your strategic planning?

- It does not.
- Dependent upon interest and funding opportunities.
- It is rarely discussed.
- Major component.
- It does not have a high profile compared to things like nano, bio, etc. Still a necessary part of the education for our alumni in engineering companies.
- It is not a critical component. Corrosion is one of the many design considerations and is equivalently important to other materials selection and materials design factors.
- Unclear at this point.
- Seen as crucial in priority areas such as materials for energy applications, nanotechnology.
- We will continue to offer a course in corrosion/electrochemistry.
- We feel it is important in a broad context that includes other failure mechanisms like wear.
- Only as an outcome for the undergraduate chemical engineering.
- The department faculty feel corrosion is a critical area where students need a basic understanding
- It is not a focus area.

What are the challenges in establishing/maintaining corrosion classes?

- It is not a required class for our students and they have very few electives in their programs.
- Student interest.
- Lack of faculty interest.
- The lack of an adequate textbook that covers ALL aspects of the environmental degradation of materials. Textbooks that are exclusive to metals are

becoming less applicable and useful. Metallic corrosion is only one of many important topics.
- Faculty with sufficient knowledge of current status of field.
- The fundamentals of electrochemistry have to be taught. This topic is not well covered in core courses.
- Getting the resources to have a substantial lab experience.
- A more exciting textbook that emphasizes more modern applications would help. In general, though, the instructor for our corrosion classes is so highly regarded that students enjoy his classes. Many students from outside of the major take them as electives.
- None.
- Professors earn tenure via research, in the main, and it is not easy to do that in the corrosion area. Thus it is difficult to maintain faculty expertise in this particular subject.
- Moderate student interest and a perception that it is less exciting than making new products.
- Enrollment numbers are low.
- Good instructors.

Do you see a role for 2-year colleges in corrosion education?
- No. We have a very strong community college system, and half of our B.S. graduates came from a 2-year program at a community college, where they covered basic physics, chemistry, and math.
- Do not see a role, but we accept 2-year college graduates in all of the engineering programs.
- No
- Maybe a role. None accepted as yet.
- In general, we would rather have students focus on foundational science and engineering at 2-year colleges.
- Yes, but we have not received applicants from any students with a corrosion background.
- Haven't seen any.

Appendix C

Publications Data

NRC staff, with the assistance of staff at the George E. Brown, Jr., Library at the National Academies and using the SCOPUS database, tracked articles in the journals *Corrosion* and *Corrosion Science* for the past 22 years. In particular, the staff tracked the following data:

- Articles per year where the lead author was U.S.-based—that is, the author's home institution was in the United States.
- Articles per year that were drafted by an author based at a higher educational facility compared with articles per year that were drafted by authors at government agencies, industry, and others.
- Number of academic institutions that had published articles compared with the number of government, industry, and others that published articles.

The data tracked are presented in Figures C-1 through C-7. These data should be taken as a sampling that may signal a larger trend.

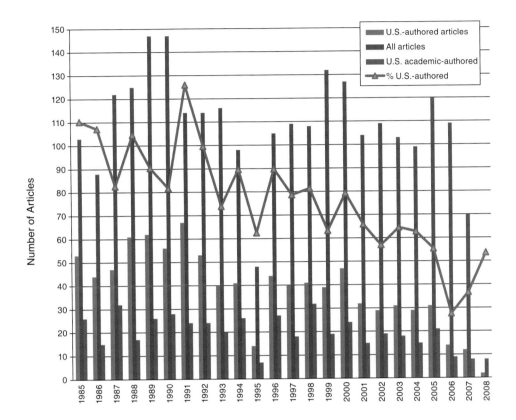

FIGURE C-1 Articles published in *Corrosion* from 1985 to 2007. The chart seems to indicate a gradual overall decline in the percentage of articles written by U.S.-based lead authors. In 1985, 52 percent of all articles in *Corrosion* had a U.S.-based lead author.

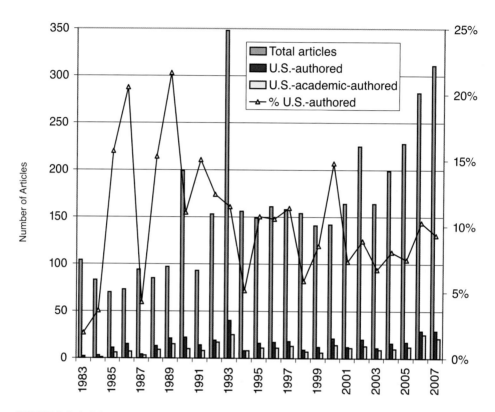

FIGURE C-2 Articles published in *Corrosion Science* from 1985 to 2007. There are generally fewer U.S.-authored articles in this publication than in *Corrosion*. The shares of articles with U.S. authors are 16 percent in 1985 and 9 percent in 2007.

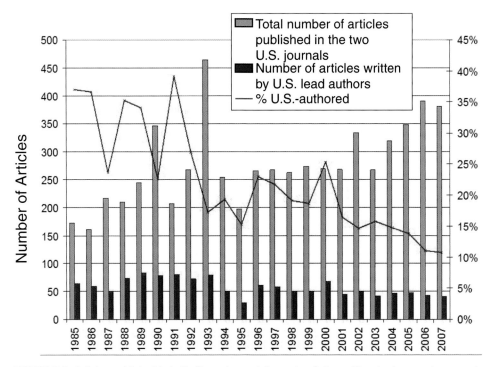

FIGURE C-3 Articles published in both *Corrosion* and *Corrosion Science*. The chart seems to suggest a gradual decline in the share of articles with a U.S. lead author relative to the total number of articles, with U.S.-authored articles having larger shares in the late 1980s and early 1990s.

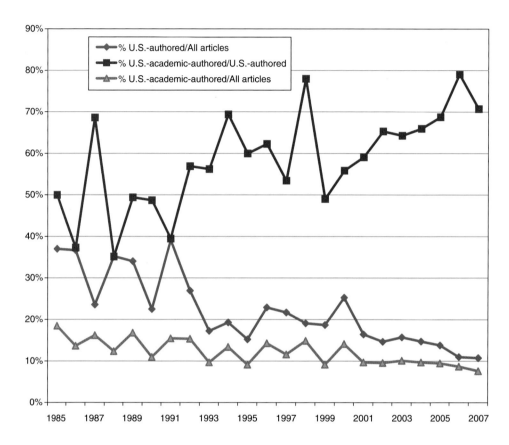

FIGURE C-4 U.S.-authored articles in *Corrosion* and *Corrosion Science* as a share of all articles and the number of articles written by U.S. academics as a share of all U.S.-authored articles and as a share of all articles in both journals. The data seem to show that while U.S.-authored articles as a share of all articles seems to decline over time, U.S.-academic-authored articles as a share of all articles seems to have declined much more slowly.

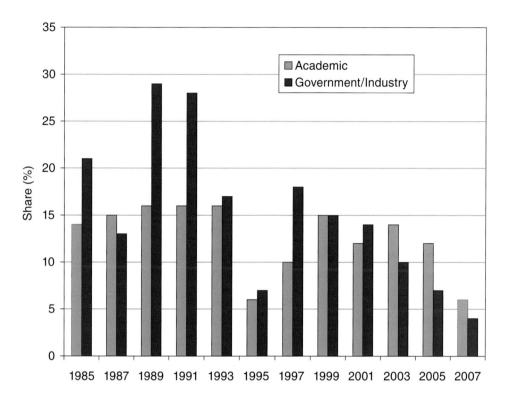

FIGURE C-5 Share, academic vs. industry and government, of U.S.-based organizations that employ authors who have published articles in *Corrosion*. Data seem to indicate that while authors with government and industry affiliations dominated until the mid-1990s, in more recent years, authors at academic organizations have been the larger source of articles.

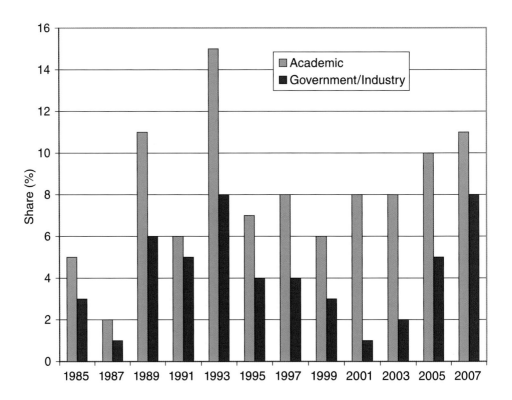

FIGURE C-6 Share, academic vs. industry and government, of U.S.-based organizations that employ authors who have published articles in *Corrosion Science*. Unlike the data for *Corrosion*, these data seem to show that academic authors have dominated the entire time.

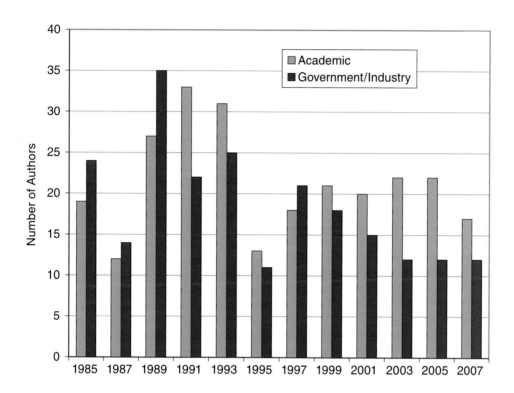

FIGURE C-7 Number of U.S. academic authors vs. U.S. government/industry authors in both journals. The data seem to indicate a spike in overall U.S. authorship in the late 1980s and early 1990s, followed by a decline. While government/industry authorship seems to have declined, academic authorship seems to be holding steady.

Appendix D

Short Courses on Corrosion

Because there was such a wide range of short courses, the committee chose to present them in a tabular format so that the reader could easily learn the focus of the course, the topics it covered, the length of the course, and the organization offering it. The courses are classified as general, platform-specific, or technology-specific.

The information in this appendix was gathered in a brief search during the summer of 2007 of Web sites, supplemented by input from committee members. The tables include most of the U.S. institutions offering continuing education and providing administrative, logistical, and curricular information about the corrosion-focused or corrosion-related onsite or online courses or modules they offer. The listings also include some certification courses that qualify individuals to be inspectors of products or processes or operators of platform- or technology-specific systems and equipment intended to prevent or mitigate degradation due to corrosion. Inclusion in these tables does not indicate committee endorsement of a course, nor does it attest to the quality of the instructor(s) or the technical content of the course.

The information is organized into three tables: courses offered by universities (Table D-1), by professional societies (Table D-2), and by private vendors (Table D-3).

TABLE D-1 University-Based Continuing Education

School	Focus	Format, Location, Length	Topics Covered
General Corrosion			
Defense Acquisition University	Corrosion awareness	100% online learning	Introduction to corrosion Planning, implementation, and management Corrosion characteristics, effects, and treatment Preventing corrosion Controlling corrosion Nonmetallic material degradation Corrosion prevention and control overview scenario Examination Survey
Ohio State University	Advanced general corrosion	Lectures and one lab, online, one quarter	Fundamentals of electrochemistry Thermodynamics of corrosion Electrochemical kinetics Corrosion laboratory Corrosion phenomenology Passivity Localized corrosion Environmentally induced cracking Coatings Inhibition Cathodic protection Atmospheric corrosion Corrosion in concrete Microbially assisted corrosion
North Dakota State University	Advanced general corrosion	70% lecture/ 30% lab, on campus, 5 days, 8 hours/day	Corrosion basics Thermodynamics and kinetics Polarization Passivation and inhibition Electrical impedance spectroscopy Types of corrosion Materials selection Cathodic protection Corrosion control by coatings

continues

TABLE D-1 Continued

School	Focus	Format, Location, Length	Topics Covered
NTU College of Engineering and Applied Science at Walden University	General corrosion	27 lectures (75 minutes each), online with videos of experiments	Electrochemical mechanism of corrosion Thermodynamics and corrosion tendency Electrode potentials and potential-pH diagram Electrochemical kinetics and corrosion rates Passivity and inhibition Theory of cathodic and anodic protection Aqueous and atmospheric corrosion Corrosion in soils, liquid metals, and fused salts Effect of mechanical stress High-temperature metal-gas reactions Materials selection Alteration of the environment Cathodic and anodic protection Metallic, inorganic and organic coatings Corrosion prediction and testing
Pennsylvania State University	Advanced general corrosion	70% lecture/ 30% lab, on campus, 5 days, 8 hours/day	Corrosion thermodynamics Corrosion kinetics polarization Rate measurement techniques Passivity/localized corrosion Electrical impedance spectroscopy Scanning probe methods in evaluation Statistical analysis of data Corrosion inhibitors
University of Virginia	Advanced general corrosion	50% lecture and 50% lab, 5 days, 8 hours/day, on campus	Introduction to corrosion Basic thermodynamics Polarization curves Passivity Pitting and crevice corrosion Test methods Resistance EI spectroscopy Cathodic protection

TABLE D-1

School	Focus	Format, Location, Length	Topics Covered
Platform-Specific			
Oklahoma State University	Pipeline	100% lecture, on campus, 2.5 days, 6 hours/day	Federal regulations Pipeline integrity overview Pre- and postassessment for gas and liquid lines Direct inspection for gas and liquid lines Inline inspection Stress corrosion cracking
	Internal track	100% lecture, on campus, 2.5 days, 6 hours/day	Federal regulations Internal corrosion mechanisms Bacteria in corrosion Laboratory and field testing Metallurgy Pigging Chemical treatment methods
	External track	100% lecture, on campus, 2.5 days, 6 hours/day	Federal regulations Electricity for cathodic protection Introduction to corrosion and cathodic protection Rectifier fundamentals Installation of conventional and deep groundbeds Field demonstration Cathodic protection instruments
University of Kansas	Aircraft structure	100% lecture, on campus and onsite, 4 days	Introduction to aircraft corrosion Mechanisms and types of corrosion Monitoring corrosion basics Corrosion control methods Detection and remediation Aircraft corrosion prevention and control Military specifications Corrosion prevention and control interpretations Current and future airplanes Aircraft maintenance procedures

continues

TABLE D-1 Continued

School	Focus	Format, Location, Length	Topics Covered
Technology-Specific			
Purdue University	Underground corrosion, cathodic protection (basic)	100% lecture, on campus, 3 days	Principles of basic electricity Forms of corrosion Corrosion prevention Cathodic protection, Part I Cathodic protection, Part II Cathodic protection equipment and applications, Part I Cathodic protection equipment and applications, Part II Cathodic protection monitoring
	Underground corrosion, cathodic protection (advanced)		Cathodic protection design—Introduction Cathodic protection design—Impressed, Part I Cathodic protection design—Impressed, Part II Cathodic protection installation Soil resistivity testing for cathodic protection installation Cathodic protection design—Galvanic, Part I Cathodic protection design—Galvanic, Part II Group problem solving
	Underground corrosion, coatings		Introduction Surface preparation Generic coating types for pipeline exteriors and interiors Coatings instrumentation Pipeline pigging Causes of coating failures Tank linings Hands-on field coating demonstration

TABLE D-2 Continuing Education Offered by Professional Societies

Organization	Focus	Format, Location, Length	Topics Covered
General Corrosion			
ASM International	General corrosion (basic)	100% lecture at ASM headquarters, 5 days	Basic concepts in corrosion The electrochemical model Corrosion kinetics and driving forces Eight major forms of corrosion Corrosion of eight alloy classes Common corrosive environments Methods of corrosion control
ASM	General corrosion (basic)	100% lecture, 15 videotaped sessions	Introduction to corrosion Basic concepts in corrosion Thermodynamics: Potential–pH diagrams Kinetics of corrosion: Polarization Eight forms of corrosion: Uniform, pitting, concentration cell Eight forms of corrosion: Galvanic, stress corrosion cracking Eight forms of corrosion: Erosion-corrosion, intergranular, dealloying Corrosion testing and monitoring Electrochemical test methods General material considerations and applications to ferrous alloys Nonferrous and nonmetallic materials Corrosive environments Economics and failure analysis Methods of control: related to coatings
ASM	Wear and corrosion	100% lecture at ASM headquarters, 5 days	Introduction Wear mechanisms Corrosion mechanisms Designing materials to match the environment Lubrication, friction, and wear testing Advanced materials design for wear and corrosion Repair strategies to optimize uptime
ASCE	Corrosion prevention and control	DVD, 3 hours	Scientific principles of corrosion Types of corrosion Practical methods for corrosion prevention

continues

TABLE D-2 Continued

Organization	Focus	Format, Location, Length	Topics Covered
NACE	Corrosion basics	Lectures and labs, 5 days	Basics of electrochemistry Types of environments Engineering materials Forms of corrosion Corrosion control and prevention methods Testing and monitoring techniques
NACE	Basic corrosion (on CD ROM)	Distance learning, self-guided, eight chapters of text, two case studies, two simulated experiments, four quizzes	Basics of electrochemistry Types of environments Engineering materials Forms of corrosion Corrosion control and prevention methods Testing and monitoring techniques
Platform-Specific			
NACE	Refining industry	Lectures and labs, 5 days	Refinery corrosion Materials for refinery applications Refinery operations and corrosion Refinery units/processes Corrosion monitoring in refineries
NACE	Pipelines	Lectures and labs, 5 days	Corrosion theory Types and mechanisms of corrosion Identification of corrosion mechanisms Investigation of pipes/components Internal corrosion mitigation Internal corrosion integrity management
NACE	Offshore	Lectures and labs, 5 days	Corrosion and corrosion control Protective coatings systems Splash-zone systems Cathodic protection systems Regulatory issues Corrosion prevention maintenance programs Facility breakdown Condition grading systems Data collection and management systems Assessment standards Safety In-service evaluation equipment Inspection planning

TABLE D-2

Organization	Focus	Format, Location, Length	Topics Covered
NACE	Shipboard corrosion	Lectures and labs, 5 days	Visual assessment exercise Corrosion theory Corrosion control Cathodic protection Protective coatings and linings Corrosion resistant materials Safety Corrosion protection system evaluation Evaluation tools and equipment
Society of Automotive Engineers	Automotive industry	100% lecture, onsite, 2 days	Insight into corrosion Underlying causes of corrosion Potential solutions for design and specific metal structures Types of corrosion Galvanic corrosion (dissimilar metals) Concentration cell corrosion Crevice corrosion Stress corrosion Corrosion-assisted fatigue Transportation vehicle design: Corrosion case histories

Certification Courses

Organization	Focus	Format, Location, Length	Topics Covered
NACE	Coating inspector program #1	6 days	Coatings introduction Curing mechanisms Role of the inspector Environmental test instruments Inspection procedures Nondestructive test instruments Coatings specifications Documentation Surface preparation and standards Application procedures Coating failures Field lab Documentation Nondestructive test and inspection

continues

TABLE D-2 Continued

Organization	Focus	Format, Location, Length	Topics Covered
NACE	Coating inspector program #2	6 days	Dehumidification Centrifugal blast cleaning WaterJetting Coating types and inspection criteria Hot-dip galvanizing Spray metalizing Concrete and cementitious surfaces Pipeline coatings Specialized application equipment Inspection instruments lab Laboratory instruments and test methods Coating survey techniques Cathodic protection
Technology-Specific			
NACE	Cathodic protection interference		Effects of stray current, A/C voltage, and telluric currents on metallic structures Detecting stray current, A/C interference and telluric current Deleterious effects of A/C and D/C interference Predicting A/C interference

TABLE D-3 Corrosion Courses Offered by Private Vendors

Organization	Focus	Format, Location, Length	Topics Covered
General Corrosion			
Corrosion College	General corrosion	Short course	Introduction Corrosion engineering: Fighting corrosion loss and damage Theory of corrosion Corrosion prevention methodology Plant operations and relevance to corrosion Applications: Consequences of real-world corrosion Corrosion prevention: Installation tools
CTC/U.S. Army Corrosion Office	Basic corrosion control	Online	Introduction to corrosion Why corrosion occurs Corrosion of specific metals Forms of corrosion Ways to control corrosion Preventive techniques
Training Technology, Inc.	Corrosion control	100% lecture, Las Vegas or onsite, 2-3 days	Introduction Corrosion basics Forms of corrosion Methods for corrosion control Identification and analysis of corrosion Metallic materials performance Nonmetallic materials performance
Platform-Specific			
Appalachian Underground Corrosion	Pipeline (basic)	100% lecture, onsite, 2.5 days, 6 hours/day	Fundamentals of corrosion Basic corrosion math Introduction to cathodic protection Pipeline electrical isolation methods Introduction to pipeline coatings Pipeline locating Potential measurements Common mistakes in cathodic protection readings Clinic: Instrumentation and measurement

continues

TABLE D-3 Continued

Organization	Focus	Format, Location, Length	Topics Covered
Appalachian Underground Corrosion (continued)	Pipeline (intermediate)		Corrosion cells in action Installation of galvanic anodes Installation of impressed current cathodic protection systems Criteria for cathodic protection Corrosion control for pipelines Impressed current interface testing Troubleshooting in cathodic protection systems Rectifier maintenance
	Pipeline (advanced)		Pipe-to-soil and surface potential analysis Materials for cathodic protection Evaluation of aboveground coating assessment techniques Stray current analysis Design of impressed current galvanic protection Design of galvanic cathodic protection Introduction to microbial protection
	Coatings		Introduction Surface preparation Generic types of coatings: Pipelines, exterior and interior Coatings instrumentation Pipeline pigging Coating failure causes Tank linings Hands-on field coatings demonstration
	Water and wastewater		Underground corrosion in water and wastewater Water quality issues for internal corrosion Evaluating existing ductile iron pipe Corrosion protection by coatings in water and wastewater Corrosion control methods that extend life of existing pipeline Economics of cathodic protection External corrosion comparison: Steel and ductile iron pipe Corrosion control monitoring for water systems Facilities Protecting existing prestressed concrete cylinder pipe Instrumentation and measurement

TABLE D-3

Organization	Focus	Format, Location, Length	Topics Covered
Corrosion Clinic	Military equipment	Lecture, onsite and distance learning, 2 days	Corrosion in the defense industry Basic concepts in corrosion The nature of military service environment Different forms of corrosion in military equipment and systems Materials and processes for corrosion control and prevention
	Automotive systems	Lecture, onsite and distance learning, 2 days	Corrosion basics Forms of corrosion encountered in automotive systems Precoated steel for automotive systems Paint systems for automotive systems Topcoats in automotive systems Corrosion in different automotive components Design considerations in automotive systems Corrosion testing in automotive systems
Center for Professional Advancement	Oil and gas industry	Lecture/ workshop, onsite, 4 days	Corrosion management Types of corrosion Corrosion control materials Corrosion monitoring and inspection Corrosion failure and analysis "Corrosion database" demonstration
Corrosion Courses	Oil and gas industry	Lecture/ workshop, onsite, 3 days	Introduction to corrosion in the oil and gas industry Introduction to electrochemistry Thermodynamics of corrosion Kinetics of corrosion Corrosion testing Oil/gas corrosion Corrosion phenemonology Corrosion control Special environments

continues

TABLE D-3 Continued

Organization	Focus	Format, Location, Length	Topics Covered
PetroSkills, LLC	Oil and gas production	Lecture/ workshop, onsite, 5 days	Oil and gas facility corrosion Corrosion-resistant metals for oil and gas facilities Materials selection and control strategy for internal corrosion External conditions effecting corrosion Materials and coatings selection process Chemical inhibition selection process Monitoring methods and applications for corrosion of facilities Life-cycle considerations with reliability/ availability impacts
Western States Corrosion Seminar	Advanced corrosion		Introduction to corrosion chemistry and corrosion cells Theory and applications of internal corrosion monitoring Materials and their role in corrosion prevention Corrosion control of concrete pressure pipe Corrosion monitoring and remote monitoring methods Screening corroded piping systems using long-range ultrasound Coating failures—Causes and avoidance Underwater photo inspection of coatings and other control systems Use and misuse of galvanic anodes
Technology-Specific			
Corrosion Clinic	Materials selection	Lecture and lab, onsite and distance learning, 3 days	Corrosion and society Basic concepts in corrosion Different forms of corrosion: Mechanisms, recognition, and prevention Corrosion resistance of common metals and alloys Design against corrosion

TABLE D-3

Organization	Focus	Format, Location, Length	Topics Covered
Western States Corrosion Seminars	Fundamentals		Introduction to corrosion chemistry and corrosion cells Fundamentals of electricity and corrosion Fundamentals of cathodic protection Cathodic protection demonstration Cathodic protection interference causes Basic cathodic protection instruments and testing techniques Cathodic protection design for sacrificial and impressed current Use and misuse of permanent reference electrodes Fundamentals of protective coatings
	Intermediate		Introduction to corrosion chemistry and corrosion cells Forms of corrosion Introduction to corrosion control design Design and installation of deep anode groundbeds Understanding soil corrosivity Cathodic protection rectifier troubleshooting Coatings inspection Decoupling and overvoltage protection on cathodic protection structures Construction problems with some cathodic protection designs
	Rectifier		Theory and operation, basic cathodic protection Rectifier component design and operation Shunts Special-purpose rectifiers and their use Wiring diagrams and schematics Rectifier and cathodic protection system efficiencies and improvements General rectifier maintenance and repairs Hands-on troubleshooting

Appendix E

Agendas for Materials Forum 2007 and Committee Public Meetings

MATERIALS FORUM 2007:
CORROSION EDUCATION FOR THE 21ST CENTURY

March 30, 2007—Washington, D.C.

SESSION I: MOTIVATION Moderator, Ralph Adler, Army Research Laboratory

8:00 am	Welcome and setting the scene	Fiona Doyle, Forum Chair
8:10	Introduction of Session I participants	Gary Fischman, NMAB
8:15	Cost of corrosion	Neil Thompson, CC Technologies
8:40	DOD's mandate on corrosion	Daniel Dunmire, DOD-OSD
	DOD's corrosion and national security needs	Lewis Sloter, DOD-OSD
9:05	The need for corrosion engineering curriculum	Aziz I. Asphahani and Helena Seelinger, NACE Foundation

SESSION II: CURRENT PRACTICE Moderator, John Scully, University of
   Virginia

10:10 am	Introduction of Session II participants	Michael Moloney, NMAB
10:15	AMPTIAC ad hoc study on corrosion education	David Rose, Quanterion Solutions, Inc.
10:30	Corrosion education: materials science	Gerald Frankel, Ohio State University
10:45	Corrosion education: mechanical engineering	Matthew Begley, University of Virginia
11:00	Corrosion education: industry needs and response	Robert Schafrik, GE Aviation
11:15	Corrosion education: industry needs and response	Ramesh Sharma, Raytheon
11:30	Panel discussion	

WORKING LUNCH WITH TALK

Noon	The Challenge of Change in a Change-Resistant Environment	Luis M. Proenza, University of Akron

SESSION III: IMPLEMENTATION Moderator, Ron Latanision, Exponent, Inc.

1:15 pm	Introduction of Session III participants	Gary Fischman, NMAB
1:20	Perspectives on implementation	Ron Latanision, Exponent, Inc.
1:30	Response from panel members (2 minutes each) followed by panel discussion with audience participation	

      George Dieter, University of Maryland
      Robert Dodds, University of Illinois, Urbana-Champaign
      David Duquette, Rensselaer Polytechnic Institute
      Mark Plichta, Michigan Technical University
      Lee Saperstein, University of Missouri, Rolla
      Mark Soucek, University of Akron

SESSION IV: NEXT STEPS Moderator, Fiona Doyle, University of California, Berkeley

3:00 pm	Overview of workshop	Fiona Doyle, Forum Chair
3:15	Looking forward to the follow-on study	Wesley Harris, ACE Chair
3:15	Discussion of NRC's corrosion education study	
4:00	Adjourn	

MEETING OF THE
COMMITTEE ON ASSESSING CORROSION EDUCATION (ACE)

June 20, 2007—Washington, D.C.

Noon	MEETING CONVENES IN OPEN SESSION Lunch	
1:00 pm	Welcome and introduction of committee and guests	Wesley Harris, Chair
1:05	Sponsor presentation, DOD	Dan Dunmire, OSD-ATL
1:45	Other government perspectives	Robert Hanrahan, DOE-NNSA Joseph Carpenter, DOE-EERE Harriet Kung, DOE-DMS&E
2:45	Break	
3:05	Congressional perspective	Vickie Plunkett, House Armed Services Committee
3:35	Origin of the study	David Rose, Quanterion
4:05	Open discussion between panel and guests on presentations and implications for the study	
6:00	MEETING CONVENES IN CLOSED SESSION	

June 21, 2007

7:30 am   MEETING CONVENES IN CLOSED SESSION
            Working breakfast and discussion, committee and staff
8:15   MEETING CONVENES IN OPEN SESSION
            Recalling the corrosion education workshop
            –Some retrospective thoughts on the workshop from Fiona Doyle,
               CEWOP Chair
            –Open discussion with all committee and guests, led by
               Fiona Doyle, CEWOP Chair
10:00   Break
10:20   MEETING CONVENES IN CLOSED SESSION

MEETING OF THE
COMMITTEE ON ASSESSING CORROSION EDUCATION

September 17, 2007—Detroit, Michigan

9:45 am   MEETING CONVENES IN OPEN SESSION
            Government panel session moderator: John Scully, ACE
            Kayle Boomer, Hanford Nuclear Waste Storage Facility
            Steven Carr, Army Aviation and Missile Command
            Vincent Hock, U.S. Army Corps of Engineers
            Bill Kepler, U.S. Bureau of Reclamation
            Stephen Sharp, Virginia Transportation Research Council
12:15 pm   Working lunch
1:15   University panel session moderator: Gerald Frankel, ACE
            Jo Howze, Texas A&M University
            Alex King, Purdue University
            Ian Robertson, University of Illinois
            Subra Suresh, Massachusetts Institute of Technology
            Peter Voorhees, Northwestern University
3:45   Break
4:00   Open microphone discussion (community input)
5:00   MEETING CONVENES IN CLOSED SESSION

MEETING OF THE
COMMITTEE ON ASSESSING CORROSION EDUCATION

December 17, 2007—Irvine, California

10:00 am	MEETING CONVENES IN OPEN SESSION
	Industry panel session moderators: Gordon Bierwagen, ACE, and Gary Was, ACE
	Aziz Asphanhani, Carus Chemical Company
	William Hedges, BP
	Michael J. Maloney, Pratt & Whitney
	Robert Mroczkwski, connNtext
	Michael R. Ostermiller, General Motors
	Leslie Spain, Dominion Generation
	Darrel Untereker, Medtronic
Noon	Working lunch
1:00 pm	University panel session moderators: Gerald Frankel, ACE, and John Scully, ACE
	Reza Abbaschian, University of California, Riverside
	Robert Cottis, University of Manchester
	Anthony Luscher, Ohio State
	Matthew O'Keefe, University of Missouri, Rolla
	Lt. Michael Plumley, U.S. Coast Guard Academy
	Bob Sinclair, Stanford University
	Dan Walsh, Cal Poly, San Luis Obispo
	Ward Winer, Georgia Institute of Technology
3:00	Break
3:30	NACE/University of Akron, Sue Louscher, corrosion survey
4:30	General discussion
5:00	MEETING CONVENES IN CLOSED SESSION

Appendix F

Sample Learning Outcomes

METALLIC MATERIALS

The following could serve as a model set of outcomes for the successful student. This set is illustrative only and is presented by the committee as an example of what it envisages could be taught about corrosion to a range of engineering students and graduates. The committee does not purport that these lists of outcomes are complete. Establishing a complete and comprehensive set of outcomes would require an effort beyond the scope of this study.

Undergraduate Engineering Students in Design Disciplines
(Mechanical, Civil, Chemical, Industrial, Aeronautical)

- Be aware of general properties of classes of materials.
- Understand that properties can be affected by processing.
- Know that resources exist for aiding materials selection.
- Understand that trade-offs in properties are often required—between, for example, strength and ductility.
- Know that material properties degrade with time in an aggressive environment.
- Know that the degradation of properties must be considered in design processes.
- Know that there are different approaches to designing against corrosion.

- Know to interact with experts in materials selection and corrosion when necessary.

Engineering Students in Materials Science and Engineering Who Upon Graduation Should Be Knowledgeable in Materials Selection

- Understand how properties depend on microstructure.
- Understand how microstructure can be manipulated through processing to get the desired properties.
- Know how to select the right material for a given application based on the optimum combination of properties, including cost, availability, and manufacturability.
- Be familiar with the forms of corrosion.
- Know how to select the best material for corrosion resistance.
- Be aware of corrosion prevention and mitigation approaches.
- Know that experimental methods exist to assess and predict the rate of corrosion.
- Know to interact with experts in corrosion when necessary.

Graduate Engineering Students Specializing in Corrosion Who Upon Graduation Will Be Experts in the Field of Corrosion

- Be able to apply thermodynamics through the Nernst equation and Pourbaix diagram to predict likely equilibrium phases present.
- Be able to explain the electrochemical nature of metallic corrosion.
  —Write anodic reactions
  —Write cathodic reactions
- Be able to use experimental methods for determining corrosion current and potential, weight loss, polarization methods.
- Predict the effect of changes in solution concentration, temperature, and velocity on corrosion behavior.
- For metallic corrosion be able to describe, explain, and apply to different situations the various forms of corrosion.
  —Uniform
  —Localized
  —Environmental cracking
- For polymers be able to describe, explain, and apply to different situations the various forms of degradation.
  —Swelling
  —Solvation
  —Environmental cracking

- Given a new set of conditions, predict the corrosion reactions and the likely type(s) of corrosion.
- Be able to apply, explain, and describe various protection schemes.
  —Design and materials selection
  —Environmental changes
  —Coatings
  —Cathodic and anodic protection
  —Inhibitors
- Be able to describe applications of appropriate instrumentation for measuring changes in materials.

NONMETALLIC AND POLYMERIC MATERIALS

A similar list of outcomes can be developed for nonmetallic and polymeric materials. Prerequisites should be chemistry, physics, mathematics, materials, mechanics, thermodynamics, and design.

- Be able to predict or identify physical processes and chemical reactions that will cause environmental degradation of materials and the changes these will cause in the materials.
- Be able to explain the chemical and physical effect on materials of
  —UV radiation, oxygen, temperature, humidity, and aqueous immersion
  —Flexing, stretching, and other physical stresses
- Be able to use experimental methods for determining changes due to environmental exposure in material properties such as the following: T_g, MW, elastic modulus, gloss, color, spectral properties (IR, UV, etc.), conductivity, dielectric constant.
- Predict the effect of changes in use environment on materials performance.
- Be able to describe, explain, and apply to different situations the various forms of materials changes:
  —Uniform degradation
  —Localized degradation
  —Propagation of degradation effects
- For polymers and composites be able to describe, explain, and apply to different situations the various forms of degradation.
  —Swelling
  —Solvation
  —Environmental cracking
- Given a new set of environmental conditions, predict the chemical reactions that could degrade the materials involved and the likely ways their performance could be degraded.

- Be able to apply, explain, and describe various ways of protecting materials from their environment:
  - —Be able to examine newly developed materials for sensitivity to damage from their use environment.
  - —Be able to predict the effect of environmental changes.
  - —Be able to choose coatings and surface films for protection.
  - —Be able to use additives such as UV absorbers and free-radical traps.
- Be able to describe the application of appropriate instrumentation for measuring changes in materials.

Appendix G

Committee Biographies

Wesley L. Harris, *Chair*, is the Charles Stark Draper Professor and head of the Department of Aeronautics and Astronautics at the Massachusetts Institute of Technology. His research focuses on theoretical and experimental unsteady aerodynamics and aeroacoustics; computational fluid dynamics; and the government policy impact on procurement of high-technology systems. Earlier, he served as the associate administrator for aeronautics at NASA. He has also served as the vice president and chief administrative officer of the University of Tennessee Space Institute. Dr. Harris has served on committees of the American Institute of Aeronautics and Astronautics (AIAA), the American Helicopter Society (AHS), and the National Technical Association (NTA) and as advisor to eight colleges, universities, and institutes. Dr. Harris earned a B.S. in aerospace engineering from the University of Virginia and an M.S. and a Ph.D. in aerospace and mechanical sciences from Princeton University. He was elected fellow of the AIAA and of the AHS for personal engineering achievements, engineering education, management, and advancing cultural diversity. Dr. Harris is an outsider to corrosion engineering and as such will act as an honest broker in the committee process as well as bring wide experience in engineering and education. Dr. Harris has served as chair and member of various boards and committees of the National Research Council (NRC), the National Science Foundation (NSF), the U.S. Army Science Board, and several state governments. He is a current member of the NRC Division on Engineering and Physical Sciences committee, the NAE panel on grand challenges in engineering, and the NAE Committee on Engineering Education.

Ralph Adler is a research metallurgist on the coatings and corrosion team in the Materials Applications Branch of the Army Research Laboratory (ARL). He works in the Weapons and Materials Research Directorate of ARL, which is located at Aberdeen Proving Ground, Maryland, and has been a federal employee working for the Army since 1985, when he joined the Army Materials Technology Laboratory. Dr. Adler is a research scientist and is not involved in the formulation or implementation of DOD policy. His research interests include science education and awareness for all levels of our society. He has taught graduate-level courses in x-ray diffraction at Northeastern University. To encourage technical collaboration between ARL and universities, he has worked with university faculty in an advisory capacity to provide project oversight and to share resources and materials as well as collaboratively by coauthoring joint publications. He has served as a science fair judge at both local and national levels and organized the initial student poster session at the 2005 Tri-Service Corrosion Conference. To improve science awareness of public school students and the quality of science education in his hometown, he has been a member of the Committee on Science Education of the Citizens for Wellesley Public Education; its charter was to enhance and enrich science education of public school students through strong support of the Wellesley public schools science faculty and by obtaining donations of scientific equipment. Dr. Adler has represented the Army on several high-level DOD panels: currently as an ARL member of the Corrosion Forum, where he is chair of the Corrosion Education Consortium; earlier as chair of Subpanel 8 (materials processing/manufacturing research) for Project RELIANCE-Technical Panel of Advanced Materials; as a member of the Laboratory Infrastructures Consolidation Study in the office of the Secretary of Defense, the JDL-TPAM Manufacturing Sciences Working Group; two sessions of the Technical Managers Acquisition Workshop; and as secretary for the Metals Panel TP-1 of the Materials Technology and Performance of the MAT group of TTCP. He is or has been a member of a number of professional committees: as service liaison on NRC panels NMAB-444 and -467; as a member of the ASM/Advisory Technical Awareness Council; on thesis review panels for WPI and Northwestern University; and played leadership roles as chair and executive committee member of the Boston section of The Metallurgical Society (TMS)/AIME. He has also participated in many Army/DOD/NSF Source Selection panels and has been an invited member of the ARL/ARO and IRAD technical review boards. Dr. Adler earned a D.Eng. in metallurgy from Yale University and has over 40 years of experience in leading and conducting sponsored or in-house research on a variety of materials science and engineering programs in both industrial and Army organizations. With his expertise in synthesis, metals processing, and materials characterization, he has authored publications and holds U.S. patents in a variety of technical areas with commercial and military applications.

Gordon P. Bierwagen is a professor of coatings and polymeric materials (CPM) at North Dakota State University (NDSU). He has a B.S. in chemistry and mathematics from Valparaiso University and a Ph.D. in physical chemistry from Iowa State University. He has been involved in R&D in coatings since his first nonacademic work at Battelle Memorial Institute in 1969. He worked in the Battelle Electrochemical Engineering Department on a project for PPG on the electrochemical characterization of anionic electrodeposition coatings and the modeling of solution changes during electrodeposition. After leaving Battelle in 1970, Dr. Bierwagen joined the Paint Research Department at the Sherwin-Williams Research Center in Chicago. In 1971, he joined the Federation of Societies for Coatings Technology (FSCT), becoming an active member of the Chicago Society. At NDSU he is still active in the FSCT, serving on its Roon Award Committee, interacting with the Coatings Industry Education Fund on research awards and undergraduate scholarships for CPM at NDSU, and attending paint shows with regularity. He has also served on the American Chemical Society (ACS) Polymeric Materials Science and Engineering (PMSE) Division's Tess Award Committee and its Program Committee and organized and published the ACS-PMSE Symposium "Corrosion and Its Control by Coatings" from the 1996 ACS national meeting in New Orleans. Since joining the NDSU CPM Department, Dr. Bierwagen has continued his research interests in the physical chemistry of coatings, foaming in W/R polymer solutions, adsorption from solution onto pigment particles, CPVC-related phenomena, and computer-based coatings design. He also set up a new research program at NDSU on corrosion control by coatings that has been very successful in attracting government and industry support. He is also working on new theoretical analyses and interpretation of CPVC-related behavior in coatings and films, corrosion protection of outdoor artistic and historical bronze sculpture, lithium-polymer battery studies (polymer/composite electrode design), opacity and film imperfections, and durability/lifetime prediction for coatings. Since joining NDSU, Dr. Bierwagen has had 105 publications and has made numerous presentations for the FSCT, NACE, the Electrochemical Society, ACS-PMSE, the Society for Protective Coatings, the American Institute of Chemical Engineers, and the American Institute of Conservation. He has made two invited presentations at the Gordon Research Conference on Coatings and was a session chair on aqueous corrosion. He was the plenary lecturer at the FATIPEC 2000 Congress in Turin, Italy, and an invited keynote speaker at EIS 2001. In 2007 he was honored with the Matiello Award of the FSCT and gave the Matiello lecture in October 2007 at the annual meeting in Toronto.

Dianne Chong is the director of materials and process technology for the Boeing Commercial Airplane Company. She is responsible for new technology development and transitioning onto Boeing products, functional excellence, and program support throughout the life cycle of the airplanes. Earlier, Dr. Chong was the head of strategic operations and business for engineering, in which capacity she is responsible for defining and implementing a solid strategy for all Boeing Engineering and for leading both the enterprise and Integrated Defense Systems engineering process councils, which cover 45,000 engineers. Before that, she was the director of materials and process technology for the Phantom Works and was responsible for development and technology transfer. In this capacity, Dr. Chong supports all Boeing business units. She was also responsible for the functional management of engineers who were matrixed to program support, production, and research areas. She was manager of materials and processes, liaison, and process control engineers who supported the fabrication centers and the production aircraft programs (F/A-18C/D, F-15, AV-8B, T45TS, and C-17). She was team leader of material and product form engineering in production aircraft programs since June 1995. Dr. Chong received a B.S. in biology and psychology (1971), an M.S. in physiology (1975), and an M.S. in metallurgical engineering (1983) from the University of Illinois. She utilized the knowledge from these two disciplines to develop porous titanium prostheses. In 1986, she completed her doctoral work by developing a steelmaking technology derived from methods used by the ancient Chinese and received her Ph.D. from the University of Illinois. She completed the degree Executive Master of Manufacturing Management at Washington University in 1998. Dr. Chong is a fellow of ASM International and a member of TMS, AIAA, SAE, SME, SWE, Beta Gamma Sigma, and Tau Beta Pi. She was a 2001 graduate of Leadership America, a 1999 participant in the Greater Missouri Leadership Challenge, and 1997 recipient of the YWCA Special Leadership Award in Science and Technology. She has received the Boeing Corporate Diversity Award (2003), the Women of Color Technology All-Star (2002), the OCA Corporate Achievement Award (2002), the Diversity Change Agent Award (2004), and the University of Illinois Alumna of the Year (2006). She is a member of the NRC National Materials Advisory Board.

George E. Dieter is the emeritus professor of mechanical engineering and the Glenn L. Martin Institute Professor of Engineering at the University of Maryland, having retired as dean of the College of Engineering in 1994. Before that, Dr. Dieter was professor of engineering and director of the Processing Research Institute at Carnegie Mellon, as well as chair of metallurgical engineering at Drexel University. He started his career at the Engineering Research Laboratory of the DuPont Company. His teaching and research interests are engineering design, materials processing, and quality engineering. Dr. Dieter is a member of the National Academy of Engineering and a fellow of AAAS, ASM International, TMS, and the American Society for Engineering Education (ASEE). He was national president of ASEE and received the Lamme Medal, its highest honor. His book *Mechanical Metallurgy* has been in print since 1961 in various editions, and his book *Engineering Design: A*

Materials and Processing Approach is in its third edition (2000). He was the editor of volume 20 of the ASM Handbook, *Materials Selection and Design*, published in 1997. He has been active on many NRC committees, including the National Materials Advisory Board. Dr. Dieter received a bachelor's degree in metallurgical engineering from the Drexel Institute of Technology and a Sc.D. from the Carnegie Institute of Technology (Carnegie Mellon).

Fiona M. Doyle is a professor in the Department of Materials Science and Engineering at the University of California at Berkeley (UCB). She is also the executive associate dean and associate dean for academic affairs in the UCB College of Engineering. She obtained a bachelor's degree in metallurgy and materials science from the University of Cambridge, England, and an M.Sc. in extractive metallurgy and a Ph.D. in hydrometallurgy from Imperial College, University of London. Dr. Doyle's main area of research is the solution processing of minerals and materials. She studies processes such as the leaching and transformation of minerals, solvent extraction, organic-phase reactions, hydrolysis, precipitation, crystallization, and electrochemical reactions from a fundamental thermodynamic and kinetic perspective. Much of her work aims to adapt the techniques used in the primary production of commodity minerals and metals for the commercial-scale processing of value-added, advanced materials. She is also engaged in ongoing research on improving the environmental impact and energy utilization associated with the production of minerals and materials. Dr. Doyle has served the state of California in assessing the environmental impact of mining and mineral processing operations and in developing policies for addressing environmental damage attributable to historic mining activities. Dr. Doyle is a member of the National Materials Advisory Board and also served as the chair of the panel that organied the Corrosion Education Workshop.

David J. Duquette received his Ph.D. in materials science from the Massachusetts Institute of Technology in 1968. Following his postgraduate work, he performed research on elevated temperature materials, joining the faculty of the Rensselaer Polytechnical Institute in 1970. He is the author or coauthor of more than 160 scientific publications, primarily in the areas of environmental degradation of materials and electrochemical processing of semiconductor interconnects. He is a recipient of NACE's Whitney Award for his contributions to corrosion science and the Alexander von Humboldt Senior Scientist Award. He is a fellow of ASM International and of NACE International. Professor Duquette's research interests include the physical, chemical, and mechanical properties of metals and alloys, with special reference to studies of environmental interactions. Current projects include studies of aqueous and elevated-temperature corrosion phenomena, the effects of corrosive environments on fatigue behavior, the environmental cracking of alloys, the role of corrosion science in understanding the planarization of metal interconnects on semiconductor devices, and the electrodeposition of semiconductor interconnects. A fundamental understanding of material–environment interactions is critical to the engineering application of metallic materials. Dr. Duquette was a member of the Panel on Electrochemical Corrosion, which has completed its work.

Gerald S. Frankel is a professor of materials science and engineering at the Ohio State University and director of the Fontana Corrosion Center. He earned an Sc.B. in materials science engineering from Brown University and an Sc.D. in materials science and engineering from the Massachusetts Institute of Technology. Before joining OSU in 1995, Dr. Frankel was a postdoctoral researcher at the Swiss Federal Technical Institute in Zurich, Switzerland, and then a research staff member at the IBM Watson Research Center. He has more than 180 publications, and his primary research interests are the passivation and localized corrosion of metals and alloys, corrosion inhibition, and protective coatings. Dr. Frankel is past chairman of the Corrosion Division of the Electrochemical Society, past chairman of the Research Committee of NACE, and a member of the editorial board of the journal *Corrosion*. He is a fellow of NACE International, the Electrochemical Society, and ASM International. He has received the Alexander von Humboldt Foundation Research Award for Senior U.S. Scientists, the H.H. Uhlig Educators Award from NACE, and the Harrison Faculty Award and Lumley Research Award from the OSU College of Engineering. In 2005 he was on sabbatical at the Max Planck Institute for Iron Research in Düsseldorf, Germany. Dr. Frankel is a member of the organizing panel of NRC's Corrosion Education Workshop.

Richard B. Griffin has been a faculty member at Texas A&M University since 1977. He earned a B.S. from Pennsylvania State University in metallurgy/metallurgical engineering and a Ph.D. from Iowa State University in metallurgy. His expertise is in materials, where he has taught and done research for more than 30 years. Dr. Griffin has worked in various areas of corrosion: erosion/corrosion, scc cracking of high-strength steels, and corrosion under organic coatings. He has also worked in tribology, where he studied the compound wear process. For almost a decade, he was a member of the Foundation Coalition team, which developed and implemented freshmen and sophomore engineering programs at Texas A&M University. Recently, Dr. Griffin helped in the establishment of a branch campus of Texas A&M University in Doha, Qatar. He has received the Texas A&M University Association of Former Students Faculty Distinguished Achievement Award for student relations and the NACE Technical Achievement Award. He is a member ASM, ASEE, and NACE.

Sylvia M. Johnson is chief of the Thermal Protection Materials and Systems Branch at NASA Ames Research Center. She holds a B.Sc. (honors) in ceramic engineering from the University of New South Wales, Australia, and an M.S. and a Ph.D. in materials science and engineering from the University of California at Berkeley. She was at SRI International in Menlo Park, California, for 18 years, where she participated in and led a wide variety of projects for government and commercial clients, both domestic and international, becoming director of ceramic and chemical product development there before leaving to join NASA. A fellow of the American Ceramic Society since 1992, Dr. Johnson served as its vice president in 1996-1997 and as an elected board member from 2002 to 2005. In addition to many committee assignments, she has been counselor of the Northern California section since 1988, chaired five Pacific Coast regional meetings, and is currently U.S. representative to the International Ceramic Federation. From 1997 to 2002, Dr. Johnson served on the National Materials Advisory Board. During that time, she chaired two NMAB materials forums and was chair of the NMAB Workshop on Education and Workforce in Materials Science and Engineering. She has served on the National Institute of Standards and Technology evaluation board and on the Assessment Panel for Materials; is on the organizing committee of the National Space and Missile Materials Symposium; is a member of NRC committees; and currently serves on the evaluation board for materials science and technology at the Sandia National Laboratories. She holds six U.S. patents, and her 55+ publications are referenced in many journals, patents, and books.

Frank E. Karasz obtained a Ph.D. in physical chemistry from the University of Washington and a B.Sc. from Imperial College, London. He spent some time at the U.K. National Physical Laboratory and at General Electric Research Laboratory, Schenectady, before joining the newly established Polymer Science and Engineering Department of the University of Massachusetts in 1967. During his years at UMass Dr. Karasz served as codirector of the NSF Materials Research Laboratory and as director of the Center for Advanced Structural and Electronic Polymers. He is currently the Silvio O. Conte Distinguished Professor Emeritus at the University of Massachusetts. His research activities are concentrated in three areas of polymer physics and chemistry: (1) polymer-polymer interactions in binary amorphous and amorphous crystalline blend systems: effects of copolymerization and microstructure; (2) computer simulations of polymer-polymer miscibility; and (3) quasi-elastic light scattering from macromolecular solutions. Dr. Karasz has more than 570 publications and is the recipient of several national and international awards. He was elected to the NAE in 1991 and is also a member of three foreign academies. Dr. Karasz is a past member of the National Materials Advisory Board (NMAB) and has served on past NRC committees including the NMAB's Panel on Functional Organic and Hybrid Materials for the Committee on Materials Research for Defense-After-Next.

Ronald M. Latanision is professor emeritus of materials science and engineering and nuclear engineering at the Massachusetts Institute of Technology and corporate vice president of Exponent, Inc. He is the author or coauthor of more than 200 scientific publications, is founder and cochair of the New England Science Teachers, and is a member of the National Academy of Engineering and the American Academy of Arts and Sciences. He has been a consultant to industry and government and has been active in organizing international conferences. He was appointed to the Nuclear Waste Technical Review Board by President Bush. Dr. Latanision received a B.S. in metallurgy from the Pennsylvania State University and a Ph.D. in metallurgical engineering from the Ohio State University. During a sabbatical in 1982-1983, he served as a science advisor to the U.S. House of Representatives Committee on Science and Technology. He has served on a number of committees at the National Academies, including several committees on science education, and he also served on the Center for Education advisory board. He is a member of the organizing committee for the NRC's Corrosion Education Workshop and the Committee on Teacher Preparation Programs in the United States. He was a member of the now inactive Committee on Undergraduate Science Education.

Glenn N. Pfendt is the general manager for the Protective Coatings Division of A.O. Smith Corporation. A University of Illinois ceramic engineer, Dr. Pfendt has been an ACerS member for over 30 years and an active ECD committee participant for 20 years, serving as chair and more recently as division trustee. An ACerS fellow, a Mueller Award winner, and a member of NACE, Dr. Pfendt serves on the boards of the Associated Industries of Kentucky, the Boone County High School, and the Porcelain Enamel Institute, where he is currently vice president. His business experience includes 15 years in his current position, where he is responsible for the division's providing technical support on coatings, materials, equipment, and processes for the application and performance of coatings as well as their design, manufacture, and global sales both within A.O. Smith as well as to external customers, primarily in the appliance, welding, and primary metals industries. Specific product developments include multiphased coatings with hot water performance two to four times greater than was formerly available in the water heater market. Prior to A.O. Smith, Dr. Pfendt conceived, built, owned, and operated Ramtec Incorporated (Rapid Melting Technologies), a company that developed, manufactured, and sold materials and coatings for the welding and joining, glass decorating, and porcelain enamel industries. In addition to joint or team developments that included a moving-electrode, continually stirred, ton-per-hour rapid-melting glass reactor that he built and the development of lead-free glass coatings for decorating tumblers, Dr. Pfendt was the sole developer of 24 new products that went to market over a 5-year period. His involvement in start-up companies and facilities includes three new ceramic product development laboratories and three coating and materials manufacturing facilities. Additional project work includes responsibility for equipment and start-up of several process lines in existing facilities, including sole design-and-build responsibilities for a flux melting, grinding, and processing plant in Mexico.

Lee W. Saperstein is dean emeritus of the School of Mines and Metallurgy and professor emeritus of mining engineering at the University of Missouri, Rolla (UMR); he served as dean from July 1, 1993, to June 30, 2004, retiring from the university at the end of December 2006. He has a B.S. in mining engineering from Montana School of Mines (now Montana Tech of the University of Montana) and a D.Phil. in engineering science from Oxford University, which he attended as a Rhodes Scholar. He was a faculty member in mining engineering at Penn State from 1967 to 1987 and for the following 6 years at the University of Kentucky, where he was also chair of the Department of Mining Engineering. While at Kentucky, he participated in an interim management team for the University of Kentucky Center for Applied Energy Research, where he was assistant director for clean coal fuels. His research specialization has been in environmental engineering of mines. He has published papers, proceeding articles, book chapters, and informal articles on this subject. He created Penn State's first surface-mining design course and supervised training programs for miners, including health and safety training and job-skills training. He is a distinguished member of his professional society, the Society for Mining, Metallurgy, and Exploration (SME-AIME), which gave him the Ivan B. Rahn Award for Education. Dr. Saperstein served as the president of ABET in 1999-2000, and he has been representative director for SME-AIME as well as secretary, president, and past president of the board of directors. He served as chair, 1989-1990, of the Engineering Accreditation Commission (EAC), where as commissioner, he led evaluation teams to 13 universities. Before he was EAC chair, he chaired the Criteria Committee when it devised the concept of "engineering topics" and wrote the first references to "program objectives" and "outcome assessments." A member of ABET's Strategic Planning Committee, he was named a fellow of ABET. He most recently served as chair of the ad hoc task force on governance, which has delivered a new constitution, bylaws, and rules of procedure to ABET. He is a holder of its Linton E. Grinter Distinguished Service Award. Dr. Saperstein has served on four state committees of selection for the Rhodes Scholarship and was secretary for the state of Kentucky.

John R. Scully is professor of materials science and engineering and codirector of the Center for Electrochemical Science and Engineering at the University of Virginia, which he joined in 1990. Earlier, Dr. Scully served as a senior member of the technical staff in the metallurgy department of the Sandia National Laboratories and ship materials engineer at the David W. Taylor Naval Ship Research and Development Center. Dr. Scully received B.E.S., M.S., and Ph.D. degrees in materials science and engineering from the Johns Hopkins University. His research interests focus on the relationship between materials structure and composition and their environmental degradation or corrosion properties, including hydrogen embrittlement, stress corrosion cracking, localized corrosion, and passivity. His corrosion research includes study of advanced aluminum, magnesium, titanium, ferrous-iron-based and nickel-based alloys, and stainless steels, as well as amorphous metals and intermetallic compounds. The development of methodologies for lifetime prediction of engineering materials in corrosive environments is also of interest. Dr. Scully teaches materials science classes as well as classes on corrosion and electrochemical aspects of materials science at both the graduate and undergraduate levels. He is a fellow of the Electrochemical Society and the National Association of Corrosion Engineers. He received the A.B. Campbell and H.H. Uhlig Awards from NACE, the T.P. Hoar Award from the Institute of Corrosion (U.K.), and the Francis LaQue Award from ASTM for his research in corrosion. He is a past recipient of the National Science Foundation Presidential Young Investigator Award. He is chair of the NACE awards committee and past chair of the NACE research committee. He is past chair of ASTM subcommittee G1.11 on electrochemical techniques in corrosion. He has served on the editorial boards of *Corrosion*, *Materials and Corrosion (Germany)*, and *Metallurgical and Material Transactions*. He served as a technical consultant to the Space Shuttle Columbia Accident Investigation Board in 2003, was a member of the Office of the Secretary of the Defense Science Board Task Force on Corrosion Control in 2004, and was the chair and organizer of the 2004 Gordon Conference on Aqueous Corrosion. He is a member of the NRC's Corrosion Education Workshop organizing panel and the subsequent study group.

Gary S. Was received his Sc.D. from MIT in 1980. He is professor of nuclear engineering and radiological sciences and also of materials science and engineering at the University of Michigan. He is currently the director of the Michigan Memorial Phoenix Energy Institute and has held positions as associate dean of the College of Engineering and chair of the Nuclear Engineering and Radiological Sciences Department. Dr. Was's research is focused on materials for advanced nuclear energy systems and radiation materials science, including environmental effects on materials, radiation effects, and ion beam surface modification of materials and nuclear fuels. He has worked extensively in experiments and modeling of the effects of irradiation, corrosion, stress corrosion cracking, and hydrogen embrittlement on iron- and nickel-based austenitic alloys. He has led the refinement of models for radiation-induced segregation to account for composition-dependent processes, and he developed the first comprehensive thermodynamic and kinetic model for chromium carbide formation and chromium depletion in nickel-based alloys.

Most recently his group led the development of proton irradiation as a technique for emulating neutron irradiation effects in reactor structural materials and has conducted some of the first stress-corrosion-cracking experiments of austenitic and ferritic alloys in supercritical water. During his tenure at the University of Michigan, Dr. Was graduated 22 Ph.D. students, created three new graduate-level courses dealing primarily with irradiation effects on materials and on nuclear fuels, and an engineering summer course on ion beam modification of materials. He served as chair of the Organization of Nuclear Engineering Department Heads and coauthored the first American Society of Engineering Education report on man-power in the nuclear industry. He is currently serving on the board of directors of the Engineering Research Council of ASEE. He has helped to organize more than a dozen technical symposia and is a member of the ASEE, the Materials Research Society, the American Society for Metals, the Minerals, Metals and Materials Society, NACE, Sigma Xi, and Tau Beta Pi. He was chair of the Materials Research Society's fall 1994 meeting. He is director of four laboratories at the University of Michigan: the Michigan Ion Beam Laboratory for Surface Modification and Analysis, the High Temperature Corrosion Laboratory, the Irradiated Materials Testing Laboratory, and the Materials Preparation Laboratory. Dr. Was has received the Presidential Young Investigator award from NSF and the Excellence in Research Award from the College of Engineering. In 2000, he was honored with the Champion H. Matthews Award from TMS and in 2004 he was awarded the Outstanding Achievement Award by the Materials Science and Technology Division of the American Nuclear Society. He is a fellow of ASM International, NACE International, and the American Nuclear Society. Dr. Was has published over 150 technical articles in refereed, archival journals, has presented over 200 conference papers, and has given more than 65 invited seminars and talks.